Environmental, Health, and Safety Portable Handbook

Environmental, Health, and Safety Portable Handbook

Gayle Woodside

McGraw-Hill

New York   San Francisco   Washington, D.C.   Auckland   Bogotá
Caracas   Lisbon   London   Madrid   Mexico City   Milan
Montreal   New Delhi   San Juan   Singapore
Sydney   Tokyo   Toronto

Library of Congress Cataloging-in-Publication Data

Woodside, Gayle.
　Environmental, health, and safety portable handbook / Gayle Woodside.
　　p.　　cm.
　Includes index.
　ISBN 0-07-071848-2 (alk. paper)
　1. Environmental engineering—Handbooks, manuals, etc.
2. Industrial hygiene—Handbooks, manuals, etc.　3. Industrial
safety—Handbooks, manuals, etc.　I. Title.
TD145.W6697　1998
626—dc21　　　　　　　　　　　　　　　　　　　　　　98-25959
　　　　　　　　　　　　　　　　　　　　　　　　　　　　CIP

McGraw-Hill

A Division of The McGraw·Hill Companies

1 2 3 4 5 6 7 8 9 0　DOC/DOC　9 0 3 2 1 0 9 8

*The sponsoring editor for this book was Robert Esposito and the
production supervisor was Tina Cameron. It was set in Times
Ten by North Market Street Graphics.*

Printed and bound by R. R. Donnelley & Sons Company.

This book is printed on recycled, acid-free paper containing a
minimum of 50% recycled, de-inked fiber.

Contents

Introduction

Environmental, Health, and Safety Portable Handbook is a composite of definitions, conversion factors, equations, checklists, tables, and information that is used most often by the environmental, health, and safety (EHS) professional in the workplace. The book should be used as a personal tool and is designed to be carried from the office to the manufacturing floor, to environmental treatment units, to the utility plant, to contractor operations, and anywhere else that the EHS expert is needed. Blank pages associated with each topic are included for personal notes and observations.

The EHS professional is responsible for multiple disciplines—including EHS program development and implementation, facility assessments, systems and process evaluation and design, behavioral assessment and management, and more. Because of the varied information required by the EHS professional, we have formatted the book in such a way as to make the information easy to find. In addition, because the book is meant to be portable, we have presented each topic in a very condensed fashion. Detailed background information and examples typically are not included.

Thus, this book is not meant to be (and cannot be) a substitute for other technical EHS books that contain specialized information about each topic addressed in this portable handbook. The competent EHS professional will have a library of engineering and technical books that he or she should refer to when making critical EHS decisions. This book, however, does provide basic information that is used on an everyday basis, and we hope that you find it useful as you perform the many tasks required of you as an EHS professional.

Acronyms

API	American Petroleum Institute
ASTM	American Society for Testing and Materials
BAT	Best Available Technology
BCT	Best Conventional Technology
BEI	biological exposure index
BMP	Best Management Practices
BPJ	Best Professional Judgement
BPT	Best Practicable Technology
CAAA	Clean Air Act Amendments
CERCLA	Comprehensive Emergency Response, Compensation, and Liability Act
CPC	chemical protective clothing
DfE	design for environment
DOT	Department of Transportation
ECD	electrostatic capture device
EPA	Environmental Protection Agency
FID	flame ionization detector
GC	gas chromatograph
HAZMAT	hazardous materials
HAZWOPER	Hazardous Waste Operations and Emergency Response
hp	horsepower
HPLC	high performance liquid chromotography
IDLH	immediately dangerous to life and health
LC	lethal concentration

LD	lethal dose
LEL	lower explosive limit
LF	linear foot
LFL	lower flammable limit
NSPS	New Source Performance Standard
MS	mass spectrometer
OSHA	Occupational Safety and Health Administration
CPE	chemical protective clothing
P2	pollution prevention
PCB	polychlorinated biphenyl
PCFID	preconcentration and flame ionization detector
PEL	permissible exposure limit
POTW	publicly-owned treatment works
PPA	Pollution Prevention Act
PPE	personal protective equipment
PSES	Pretreatment Standards for Existing Sources
psi	pounds per square inch
PSM	process safety management
PSNS	Pretreatment Standards for New Sources
PUF	polyurethane foam
RCRA	Resource Conservation and Recovery Act
SCBA	self contained breathing apparatus
std	standard
TLV-TWA	threshold limit value-time weighted average
TSD	treatment, storage, and disposal
UEF	upper explosive limit
UFL	upper flammable limit
UST	underground storage tank
UV	ultraviolet

Environmental, Health, and Safety Portable Handbook

PART 1

General

NOTES

A. Regulations and Voluntary Initiatives Pertaining to Environmental, Health, and Safety (EHS)

1. Focus of Key EHS Regulations from 1969 to 1997 (See Table 1.A.1.)

TABLE 1.A.1 EHS Regulations

Time Frame	Key Regulations Enacted	Principle Focus Areas
1969–1975	National Environmental Policy Act (NEPA); Occupational Safety and Health Act (OSHA); Clean Air Act (CAA); Federal Water Pollution Control Act (FWPCA); Hazardous Materials Transportation Act; Energy Reorganization Act.	Cleanup of rivers and air; establishment of priority and criteria pollutants; worker protection; safe transport of hazardous materials; division of authority of defense and civilian nuclear energy activities.
1976–1983	Toxic Substance Control Act (TSCA); Resource Conservation and Recovery Act (RCRA); CAA Amendments; Clean Water Act (CWA); Mine Safety and Health Act (MSHA); Comprehensive Environmental Response, Compensation, and Liability Act (CERCLA); Uranium Mill Tailings Radiation Control Act; Low-Level Radiation Waste Policy Act; Nuclear Waste Policy Act.	Registration of hazardous materials; hazardous waste management; cleanup of land and groundwater; additional requirements for air and water; safety in mining activities; reporting of spills or unpermitted releases to the environment; regulation of uranium and thorium mill tailings; handling of low-level radiation waste on state level; disposal of civilian spent fuel and high-level waste.

TABLE 1.A.1 **EHS Regulations** (*Continued*)

Time Frame	Key Regulations Enacted	Principle Focus Areas
1984–1989	Hazardous and Solid Waste Amendments (RCRA); Emergency Planning and Community Right to Know (CERCLA); Water Quality Act; Hazard Communication Standard (OSHA).	Additional requirements for facilities managing hazardous waste; annual chemical release reporting; control of toxic pollutants to waterways; knowledge of hazards for worker and community.
1990–1997	Pollution Prevention Act (PPA), Clean Air Act Amendments (CAAA); Hazardous Waste Operations and Emergency Response (OSHA); Control of Hazardous Energy (OSHA); Standard for Process Safety Management of Highly Hazardous Chemicals (OSHA); Standard for Occupational Exposure to Bloodborne Pathogens (OSHA); Standard for Permit-Required Spaces (OSHA); Update of Standard for Personal Protective Equipment (OSHA); Update of Standard for Asbestos (OSHA).	Proactive programs with latitude for developing best technological practices in waste management; areas of United States with worst air quality problems identified and remedies legislated; emphasis on training and hazard prevention; emphasis on worker safety.

___ 2. Index of key regulations

    ___ *a. Title 10—Energy*

        ___ Parts 0 to 199, Nuclear Regulatory Commission

            ___ Part 10—Notices, instructions, and reports to workers

            ___ Part 20—Standards for protection against radiation

___ Parts 30 to 55—Licensing
___ Part 60—Disposal of high-level radioactive wastes in geologic repositories
___ Part 61—Licensing requirements for land disposal of radioactive waste
___ Part 62—Criteria and procedures for emergency access to nonfederal and regional low-level waste disposal facilities
___ Part 70—Domestic licensing of special nuclear material
___ Part 71—Packaging and transportation of radioactive material
___ Part 72—Licensing requirements for the independent storage of spent nuclear fuel and high-level radioactive waste
___ Parts 200 to 999, Department of Energy
___ b. *Title 29—Labor*
___ (1) Parts 1900 to 1999, Occupational Safety and Health Administration (OSHA)
___ (2) Part 1910—Occupational safety and health standards
___ Subpart A—General
___ Subpart B—Adoption and extension of established federal standards
___ Subpart C—General safety and health provisions
___ Subpart D—Walking-working surfaces
___ Subpart E—Means of egress
___ Subpart F—Powered platforms, manlifts, and vehicle-mounted work platforms
___ Subpart G—Occupational health and environmental control

___ Subpart H—Hazardous materials
___ Subpart I—Personal protective equipment
___ Subpart J—General environmental controls
___ Subpart K—Medical and first aid
___ Subpart L—Fire protection
___ Subpart M—Compressed gas and compressed air equipment
___ Subpart N—Materials handling and storage
___ Subpart O—Machinery and machine guarding
___ Subpart P—Hand and portable powered tools and other hand-held equipment
___ Subpart Q—Welding, cutting, and brazing
___ Subpart R—Special industries
___ Subpart S—Electrical
___ Subpart T—Commercial diving operations
___ Subpart U–Y—Reserved
___ Subpart Z—Toxic and hazardous substances
___ (3) Part 1926—Safety and health regulations for construction
___ c. *Title 40—Environmental Protection Agency (EPA)*
___ Parts 1–149—Air programs
___ Parts 150–189—Pesticides
___ Parts 190–211—Radiation protection programs
___ Parts 220–238—Ocean dumping
___ Parts 240–299—Solid wastes
___ Parts 260–272—Management of hazardous waste

___ Parts 280–281—Underground storage tank program

___ Parts 300–399—Superfund, emergency planning, and Community Right-to-Know programs

___ Parts 400–503—Effluent guidelines and standards (including sludge)

___ Parts 700–799—Toxic Substance Control Act

___ *d. Title 49—Transportation*

___ Parts 100–110—Transportation Programs Bureau

___ Part 130—Oil spill prevention and response plans

___ Parts 171–180—Hazardous materials regulations

___ Part 171—General information, regulations, and definitions

___ Part 172—Hazardous materials tables and hazardous materials communications regulations

___ Part 173—Shippers—General requirements for shipments and packagings

___ Part 174—Carriage by rail

___ Part 175—Carriage by aircraft

___ Part 176—Carriage by vessel

___ Part 177—Carriage by public highway

___ Part 178—Specifications for packagings

___ Part 179—Specifications for tank cars

___ Part 180—Continuing qualifications and maintenance of packagings

___ 3. Voluntary initiatives

___ *a. International Organization for Standardization (ISO): ISO 14000* (See Fig. 1.A.1.)

___ Standards are divided into two major groups.

Figure 1.A.1

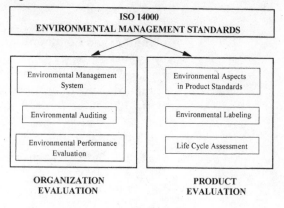

_____ Each group has three standards categories.
_____ Organization management standards are complete or near completion.
_____ Product standards are still under development.
_____ *b. ISO 14001: Environmental Management System Standard* (Ref. 44)
   _____ (1) Key elements of ISO 14001 (See Fig. 1.A.2.)
      _____ 4.1 General requirements
      _____ 4.2 Environmental policy

Figure 1.A.2

___ 4.3.1 Environmental aspects

___ 4.3.2 Legal and other requirements

___ 4.3.3 Objectives and targets

___ 4.3.4 Environmental management program

___ 4.4.1 Structure and responsibility

___ 4.4.2 Training, awareness, and competence

___ 4.4.3 Communication

___ 4.4.4 Environmental management system documentation

___ 4.4.5 Document control

___ 4.4.6 Operational control

___ 4.4.7 Emergency preparedness and response

___ 4.5.1 Monitoring and measurement

___ 4.5.2 Nonconformance and corrective and preventive action

___ 4.5.3 Records

___ 4.5.4 Environmental management system audit

___ 4.6 Management review

___ (2) Benefits of implementing ISO 14001

___ Positions the organization in the marketplace as an environmental leader

___ Provides a framework for establishing an integrated approach to environmental management that is system dependent and not person dependent

___ Promotes a positive image in the community and with governmental agencies

___ Provides a systematic, repeatable process to achieve continual improvement of the organization's environmental management system, thereby promoting improvement of environmental performance

___ Supports integration of relevant levels and functions into the environmental management system

___ Provides an internationally recognized standard for the organization to use to demonstrate to employees and interested parties its commitment to sound environmental management

___ c. *OSHA's Voluntary Protection Programs (VPP): program elements* (Ref. 43)

    ___ (1) Safety and Health Program

       ___ Documents required safety programs, such as hazard communications, medical surveillance, process safety management, laser safety, personal protective equipment, lockout/tagout, and others

       ___ Specifies safety and health training, including process equipment safety, electrical safety, confined space, training on protective equipment, and task-related training

       ___ Provides safety manual accessible to managers and employees at worksite

       ___ Provides job-specific safety plans that are kept in work area

       ___ Includes safety and health evaluation system that addresses hazard assessment, safety rules, procedures, employee involvement, and other information

       ___ Mandates that site have lower than the national industry average for injury rates and lost work days over last three years

___ (2) Assessment and control of hazards
    ___ Includes system for identifying and controlling workplace hazards
    ___ Provides program for conducting self-assessments
    ___ Documents corrective and preventive action for problems found during self-assessments
    ___ Provides documented environmental monitoring programs such as monitoring for hazardous substances, noise, and radiation
    ___ Includes proper training for personnel conducting environmental monitoring
    ___ Provides for good communication of unsafe or unhealthful work conditions
    ___ Ensures adequate medical-monitoring program

___ (3) Training
    ___ Includes general training such as good housekeeping, safe lifting, and other general safety
    ___ Includes training for management and employees, including new hires and new managers
    ___ Covers specific topics such as emergency response and site evacuation, use of protective equipment and clothing, and safe practices when using equipment

___ (4) Safety planning, work procedures, and rules
    ___ Integrates safety and health planning into overall management system
    ___ Documents safety rules and their enforcement

___ Provides for routine hazards analysis
___ Includes emergency planning
___ Defines roles and responsibilities of supervisors
___ (5) Management commitment
___ Provides for and documents top management commitment to and involvement in worker safety and health protection
___ Provides clear system of accountability and clear assignment of safety responsibilities at the line level
___ Ensures authorization of appropriate resources
___ Ensures contract workers are provided with safety and health protection equivalent to that of regular employees
___ (6) Employee participation
___ Employees are actively involved in safety and health programs.
___ Employees support safety and health programs.

NOTES

NOTES

___ B. EHS Compliance Audits

___ 1. Audit process
  ___ Review site activities and documentation before audit begins.
  ___ Conduct opening meeting.
  ___ Begin audit, per documented agenda.
  ___ Provide feedback of findings.
  ___ Conduct closing meeting.
  ___ Issue audit report.

___ 2. EHS compliance checklist (Ref. 42)
  ___ a. Permit exceptions (List all during calendar year.)
    ___ Air emissions permit limits
    ___ Wastewater/stormwater discharge permit limits
    ___ Hazardous waste treatment, storage, disposal permit
    ___ Hazardous materials storage permit requirements
    ___ Hazardous materials transportation permit requirements
    ___ Other permits/limits (list)
  ___ b. Monitoring requirements/exceptions
    ___ Air-monitoring data complete and accurate
    ___ Air-sampling and analysis methods followed
    ___ Wastewater/stormwater-monitoring data complete and accurate
    ___ Wastewater/stormwater-sampling and analysis methods followed
    ___ Waste stream profiles up to date and accurate
    ___ Personal monitoring data acquired and provided to employees, per written program or regulatory requirements
    ___ Indoor air–monitoring data acquired, per written program

___ Film badges or other radiation-monitoring data available

___ Noise monitoring set up in established or potential hearing conservation areas

___ Medical monitoring in place for employees routinely exposed to chemicals above the TLV or to meet other requirements

___ c. Incidents/accidents

___ Spills requiring emergency response

___ Spills reportable to a government agency

___ OSHA-recordable accidents

___ OSHA-reportable accidents

___ d. Administrative reports/documentation/exceptions

___ Air emissions compliance reports

___ Hazardous waste reports

___ Hazardous waste manifests

___ Hazardous waste inspection logs

___ Tier II report

___ Toxic release inventory report

___ Wastewater/stormwater compliance reports

___ PCB logs

___ OSHA 200 log

___ Confined space entry permits

___ Lockout/tagout procedures

___ Training records for HAZCOM, RCRA, DOT, PPE, confined space, PSM, HAZWOPER

___ Other training records (list)

___ Equipment inspection logs

___ Emergency response plan

___ e. Field audits/findings

___ Chemical and waste–labeling adequate

___ Chemicals segregated properly

___ Waste segregated properly and stored less than 90 days (as applicable)

___ Appropriate signs such as *Danger, Caution, No Smoking, Authorized Personnel Only, Safety Glass Area,* and others

___ Procedures being followed

___ *f.* Agency inspections/findings (List all during calendar year.)

___ Federal agency inspections

___ State agency inspections

___ Local agency inspections

___ *g.* Key EHS measurements (See Fig. 1.B.1.)

___ Hazardous waste reductions (Use base year 1988.)

___ Nonhazardous waste recycling and reduction (Include recycle tonnage by product, such as paper, plastics, and metals; also include track revenues year to year.)

___ Chemical use (by entire facility and by process center)

___ Water use (by entire facility and by process center)

___ Energy use (cost per year/person)

___ Treatment efficiency (by treatment process)

___ OSHA rate (trend by month or quarter over at least three years)

___ Worker's compensation costs (Track over several years.)

Figure 1.B.1

NOTES

__ C. Organizational Management

__ 1. Management grid (See Fig. 1.C.1.)
  __ (1,1) manager is one who has low concern for production and low interest in people.
  __ (1,9) manager is one who has a higher concern for people than for production.
  __ (9,1) manager is one who is interested more in production than in interests of employees.
  __ (9,9) manager is concerned equally with production and with people.

Figure 1.C.1

Management Grid

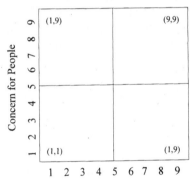

__ 2. Staffing and resources
  __ *a.* Typical professional degrees
    __ Environmental engineering
    __ Environmental science
    __ Civil engineering
    __ Chemical engineering
    __ Engineering science
    __ Safety engineering
    __ Industrial hygiene
    __ Industrial engineering

___ Chemistry
___ Natural sciences
___ *b.* Typical certifications
___ Professional engineer
___ Certified safety professional
___ Associate safety professional
___ Occupational health specialist
___ Certified industrial hygienist
___ Certified hazardous materials manager
___ Registered auditor
___ Diplomate of environmental engineering
___ *c.* Specialized training
___ RCRA training
___ DOT training
___ HAZWOPER training
___ Air quality modeling
___ Water quality modeling

NOTES

NOTES

__ D. Worker Motivation

__ 1. Maslow's Hierarchy of Needs (See Fig. 1.D.1.) (Refs. 25 and 43)

Figure 1.D.1.

__ 2. Herzberg's Theory of Motivation-Hygiene: Factors that can lead to changes in attitudes about the job (Ref. 21)

    __ *a.* Environmental—or hygiene—factors that produce short-term changes only

        __ Working conditions

        __ Wages

        __ Supervision

        __ Company policies

    __ *b.* Factors that can produce worker motivation

        __ Achievement

        __ Recognition of accomplishment

        __ Work itself

        __ Responsibility

        __ Advancement

__ 3. McGregor's Theory X and Theory Y (Ref. 26)

    __ *a.* Theory X: Conventional management beliefs about workers (circa 1960)

        __ Employees should have little freedom.

___ Employees have an inherent dislike for work.

___ Most people need to be coerced, controlled, and directed.

___ The average person wishes to avoid responsibility.

___ The average person has relatively little ambition.

___ The average person wants security above all.

___ *b.* Theory Y: McGregor's Alternative to Theory X

___ Managers should allow employees to see and reach level of self-esteem and self-actualization in their jobs.

___ Personal expenditure of physical and mental effort in work is as natural as in play.

___ Under proper conditions, persons will seek responsibility.

___ Strict supervisory control actually reduces worker effectiveness and harms employee motivation.

___ A freer environment enables workers to fully utilize their intrinsic abilities.

___ 4. Peter Drucker: Management by Objectives (Ref. 14)

___ Managers should set clear, measurable objectives.

___ Goals should be prioritized and targets set.

___ Timetables and strategies should be defined.

___ Different viewpoints and approaches should be investigated for a specific task.

___ 5. Safety management today

___ Includes integration of behavior and personal approaches

___ Avoids placing blame on individual

___ Focuses on the design of the process or activity in order to reduce accidents

___ Uses observations of behavior and the sharing of these observations with workers

___ Involves worker participation in defining programs and solving safety problems

___ Utilizes recognition and incentives for workers

NOTES

___ E. Accident Prevention

___ 1. *Job Safety Analysis (JSA)* (Ref. 32)

    ___ a. JSA is an accident prevention tool that
- ___ Works by finding hazards
- ___ Eliminates or minimizes them before the job is performed
- ___ Eliminates or minimizes them before the hazards have a chance to cause accidents

    ___ b. Use the JSA for
- ___ Job clarification and hazard awareness
- ___ As a guide in new employee training
- ___ As a refresher for jobs that are performed infrequently
- ___ As an accident investigation tool
- ___ As a means of informing employees about specific job hazards and protective measures

    ___ c. Prioritize JSAs by considering the following:
- ___ Has job produced disabling injuries?
- ___ Does job have a history of many accidents?
- ___ Does job have high potential for disabling injury or death?
- ___ Is this a new job with no accident history or evaluation?

    ___ d. Elements of JSA
- ___ (1) Sequence of basic job steps
  - ___ Break the job down into steps or tasks.
  - ___ Review all sets of movements within each task.
  - ___ Document these sets of movements.
- ___ (2) Identification of potential hazards
  - ___ Identify hazards associated with each step.
  - ___ List physical hazards and health hazards.

_____ Identify obvious hazards and hidden hazards.
_____ (3) Recommended action or procedure
    _____ Consider elimination of the hazard through engineering controls.
    _____ Consider providing personal protective equipment.
    _____ Consider the need for job instruction or training.
    _____ Review the need for housekeeping changes/enhancements.
    _____ Review the need for ergonomic changes/enhancements.
    _____ Correct serious hazards immediately.

_____ 2. *NIOSH statistics pertaining to fatal injuries—by industry division* (See Fig. 1.E.1.) (Ref. 13)

Figure 1.E.1

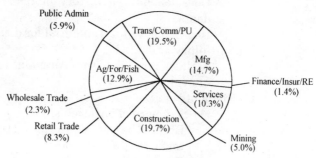

NOTES

NOTES

__ F. Risk and Incident Communication

__ 1. Risk communication to employees
     __ a. Risk communication topics
         __ (1) Chemical/waste hazard communication
            __ Basic information required as part of OSHA's Hazard Communication Standard, including hazards of chemicals in work area, chemical labeling, material safety data sheets, and other information
            __ Hazards of chemicals used on site
            __ Examples of home chemical hazards
            __ Safe storage methods for chemical waste
            __ Safe methods of handling chemicals and waste
            __ Proper protective equipment to be used when handling chemicals and waste
            __ Information about chemical/waste disposal
            __ Information about pollution prevention/recycling efforts
            __ Information about how individual employees can make the workplace safer and their activities more environmentally responsible
         __ (2) Equipment and process safety communication
            __ Communication about proper equipment use and consequences of misuse
            __ Process safety information
            __ Process hazard safeguards such as redundancies, interlocks, and fail-safe mechanisms
            __ Location and use of emergency stop buttons

__ Information about employee responsibility with respect to equipment
and process safety

__ (3) Fire and life safety communication

__ Location of exits and fire-safe corridors

__ Location of fire extinguishers

__ Emergency phone number, such as
site security or local fire/emergency
number

__ Emergency evacuation procedures

__ Emergency signals such as alarms,
sirens, or bells

__ Information about past incidents and
accidents

__ *b.* Methods of risk communication

__ "All hands" meetings where middle and
senior management present information
to line employees

__ Department meetings

__ Training of employees by EHS staff

__ Plant tours

__ Brochures, reports, and newsletters
mailed to employees

__ Training videos

__ PC-based interactive classes

__ On-the-job training

__ Test drills

__ 2. Risk communication to public and interested parties

__ *a.* Risk communication topics

__ (1) Communications about risks of chem-
icals/wastes found at facility

__ Types and amounts of chemicals
stored at the facility

__ Types and amounts of wastes generated at the facility

__ Chemical/waste safety precautions

___ Risk/occurrence of release migration off site

___ Pollution prevention efforts

___ Recycling efforts

___ (2) Communications about equipment and process safety

    ___ Information about site safety programs and training

    ___ Information about other employee/ community protection programs

    ___ Information about equipment and process safety mechanisms

___ (3) Communications about emergency planning

    ___ Information about emergency-planning arrangements

    ___ Communications about emergency drills

___ *b.* Methods of risk communication

    ___ Citizens' advisory groups

    ___ Facility tours/open houses

    ___ Regular meetings with local officials

    ___ Meetings with regulatory agency officials

    ___ Speakers bureaus

    ___ Community newsletters

    ___ Teachers' seminars

    ___ Special telephone number for citizen questions

___ *c.* Citizen "outrage" factors to consider when communicating risk to the public (Ref. 20)

    ___ Voluntary decisions versus involuntary decisions

    ___ Natural risk versus industrial risk

    ___ Benefits to those at risk versus benefits to those not at risk (fair versus unfair)

    ___ Familiar risks versus risks that are unfamiliar

___ Well-publicized perceived incidents of risk that are being embarked on again

___ Dread of certain diseases or outcomes such as cancer

___ Moral risk (e.g., war) versus immoral risk (e.g., toxic pollution discharged into waterways)

___ Quantifiable data, to show hazard prevention, versus no data

___ Situation controlled by individual versus controlled by system

___ Entity (individual) considered trustworthy versus untrustworthy

___ Decision-making process open versus process closed

___ 3. Incident communication to agencies

___ Preplan who will provide the communication.

___ Ensure the reporting time frame is met.

___ Provide your name, facility name, address, and phone number.

___ Provide description of incident, including extent of damage to human health and/or the environment, any off-site impact, or other impacts.

___ Provide information on emergency response, including site, local, state, national, or contractor response.

___ Provide information on corrective action, cleanup, and/or remediation.

___ Provide information on preventative action.

NOTES

NOTES

G. Conversion Factors

To Obtain:	Multiply:	By:
acres	Sq miles (mi^2)	640
atmospheres (atm)	cm of Hg @ 0°C	0.013158
atmospheres	ft of H_2O @ 39.2°F	0.029499
atmospheres	grams/sq cm (g/cm^2)	0.00096784
atmospheres	in Hg @ 32°F	0.033421
atmospheres	in H_2O @ 39.2°F	0.0024583
atmospheres	pounds/sq ft (lb/ft^2)	0.00047254
atmospheres	pounds/sq in (lb/in^2)	0.068046
Btu	ft-lb	0.0012854
Btu	hp-h	2545.1
Btu	kg-cal	3.9685
Btu	kWh	3413
Btu	watt-hr (Wh)	3.413
calories (cal)	ft-lb	0.32389
calories	joules (J)	0.23889
calories	Wh	860.01
centimeters (cm)	inches (in)	2.54
centimeters	microns (μm)	0.0001
centimeters	mils	0.00254
cm of Hg @ 0°C	atm	76
cm of Hg @ 0°C	ft of H_2O @ 39.2°F	2.242
cm of Hg @ 0°C	g/cm^2	0.07356
cm of Hg @ 0°C	in H_2O @ 4°C	0.1868
cm of Hg @ 0°C	(lb/in^2)	5.1715
cm of Hg @ 0°C	(lb/ft^2)	0.035913
cu cm (cm^3)	in^3	16.387
cu ft (ft^3)	m^3	35.314
cu ft	yd^3	27
cu ft	gal (USA, liq)	0.13368
cu ft	liters (l)	0.03532
cu in (in^3)	cm^3	0.061023
cu meters (m^3)	ft^3	0.028317
cu meters	yd^3	0.7646
cu meters	gal (USA, liq)	0.0037854
cu meters	l	0.001000028
cu yards (yd^3)	m^3	1.3079
feet (ft)	m	3.281
gallons (USA, liq) (gal)	barrels (bbl) (petroleum, USA)	42
gallons (USA, liq)	ft^3	7.4805
gallons (USA, liq)	m^3	264.173
gallons (USA, liq)	yd^3	202.2
gallons (USA, liq)	l	0.2642

To Obtain:	Multiply:	By:
grains (gr)	g	15.432
grains/gal (USA, liq) (gr/gal)	parts/million (ppm)	0.0584
grams (g)	gr	0.0648
grams	ounces (oz) (avoir)	28.35
grams	lb (avoir)	453.5924
inches (in)	cm	0.3937
inches	μm	0.00003937
inches	mils	0.001
in of Hg @ 32°F	atm	29.921
in of Hg @ 32°F	lb/in^2	2.036
joules (J)	Btu	1054.8
joules	cal	4.186
joules	ft-lb	1.35582
joules	kg-m	9.807
joules	kWh	3,600,000
joules	mech. hp-h	2,684,500
kg-cal	Btu	0.252
kg-cal	ft-lb	0.00032389
kg-cal	J	0.0002389
kg-cal	kWh	860.01
kg-cal	mech. hp-h	641.3
kg-cal/kg	Btu/lb	0.5556
kg-cal/kWh	Btu/kWh	0.252
kg/l	lb/gal (USA, liq)	0.11983
kg/m	lb/ft	1.488
kg/cm^2	atm	1.0332
kg/cm^2	lb/in^2	0.0703
kg/m^2	lb/ft^2	4.8824
kg/m^2	lb/in^2	703.07
km	miles (USA, statute)	1.6093
kWh	Btu	0.000293
kWh	ft-lb	0.0000003766
kWh	kg-cal	0.0011628
liters	ft^3	28.316
liters	in^3	0.01639
liters	gal (USA, liq)	3.78533
mechanical horsepower/ hour (mech. hp-h)	Btu	0.00039292
mech. hp-h	ft-lb	0.00000050505
mech. hp-h	kg-cal	0.0015593
Mech. hp-h	kWh	1.341
meters (m)	ft	0.3048

To Obtain:	Multiply:	By:
meters	in	0.0254
microns (μm)	in	25,400
microns	mils	25.4
miles (USA, statute) (mi)	km	0.6214
miles (USA, statute)	m	0.0006214
millimeters (mm)	μm	0.001
mils	cm	393.7
mils	in	1000
mils	μm	0.03937
minutes (min)	radians (rads)	3437.75
ounces (oz) (avoir)	gr (avoir)	0.0022857
ounces (avoir)	g	0.035274
ounces (USA, liq)	gal (USA, liq)	128
parts/million	gr/gal (USA, liq)	17.118
pounds (lb) (avoir)	gr	0.0001429
pounds (avoir)	g	0.0022046
pounds (avoir)	kg	2.2046
pounds/sq inch (lb/in^2)	atm	14.696
pounds/sq inch	cm of Hg @ 0°C	0.19337
pounds/sq inch	in Hg @ 32°F	0.491
quarts (qt) (USA, liq)	l	1.057
square feet	acres	43,560
square feet	m^2	10.764
square meters	acres	4046.9
square meters	ft^2	0.0929
yards	m	1.0936

NOTES

__ H. Selected Standard Industrial Classification (SIC) Codes

___ 20 Food and kindred products

___ 21 Tobacco products

___ 22 Textile mill products

___ 23 Apparel and other finished products made from fabrics and other similar materials

___ 24 Lumber and wood products, except furniture
    ___ 2411 Logging
    ___ 2421 Sawmills and planing mills, general
    ___ 2426 Hardwood dimension and flooring mills
    ___ 2429 Special product sawmills, n.e.c.*
    ___ 2431 Millwork
    ___ 2491 Wood preserving
    ___ 2493 Reconstituted wood products

___ 25 Furniture and fixtures

___ 26 Paper and allied products
    ___ 2611 Pulp mills
    ___ 2621 Paper mills
    ___ 2631 Paperboard mills

___ 27 Printing, publishing, and allied industries

___ 28 Chemicals and allied products
    ___ 2812 Alkalies and chlorine
    ___ 2813 Industrial gases
    ___ 2816 Inorganic pigments
    ___ 2819 Industrial inorganic chemicals, n.e.c.*
    ___ 2821 Plastics materials, synthetic resins, and non-vulcanizable elastomers
    ___ 2822 Synthetic rubber (vulcanizable elastomers)
    ___ 2823 Cellulosic manmade fibers
    ___ 2824 Manmade organic fibers, except cellulosic
    ___ 2833 Medicinal chemicals and botanical products
    ___ 2834 Pharmaceutical preparations
    ___ 2835 In vitro and in vivo diagnostic substances

___ 2836 Biological products, except diagnostic substances

___ 2841 Soap and other detergents, except specialty cleaners

___ 2842 Specialty cleaning, polishing, and sanitation preparations

___ 2843 Surface active agents, finishing agents, sulfonated oils, and assistants

___ 2844 Perfumes, cosmetics, and other toilet preparations

___ 2851 Paints, varnishes, lacquers, enamels, and allied products

___ 2861 Gum and wood chemicals

___ 2865 Cyclic organic crudes and intermediates, and organic dyes and pigments

___ 2869 Industrial organic chemicals, n.e.c.*

___ 2873 Nitrogenous fertilizers

___ 2874 Phosphatic fertilizers

___ 2875 Fertilizers, mixing only

___ 2879 Pesticides and agricultural chemicals, n.e.c.*

___ 2891 Adhesives and sealants

___ 2892 Explosives

___ 2893 Printing ink

___ 2895 Carbon black

___ 2899 Chemicals and chemical preparations, n.e.c.*

___ 29 Petroleum refining and related industries

___ 2911 Petroleum refining

___ 2951 Asphalt paving mixtures and blocks

___ 2952 Asphalt felts and coatings

___ 2992 Lubricating oils and greases

___ 2999 Products of petroleum and coal, n.e.c.*

___ 30 Rubber and miscellaneous plastics products

___ 31 Leather and leather products

___ 32 Stone, clay, glass, and concrete products

___ 33 Primary metal industries
  ___ 3312 Steel works, blast furnaces (including coke ovens), and rolling mills
  ___ 3313 Electrometallurgical products, except steel
  ___ 3315 Steel wiredrawing and steel nails and spikes
  ___ 3316 Cold-rolled steel sheet, strip, and bars
  ___ 3317 Steel pipe and tubes
  ___ 3321 Gray and ductile iron foundries
  ___ 3322 Malleable iron foundries
  ___ 3324 Steel investment foundries
  ___ 3325 Steel foundries, n.e.c.*
  ___ 3331 Primary smelting and refining of copper
  ___ 3334 Primary production of aluminum
  ___ 3339 Primary smelting and refining of nonferrous metals, except copper and aluminum
  ___ 3341 Secondary smelting and refining of nonferrous metals
  ___ 3351 Rolling, drawing, and extruding of copper
  ___ 3353 Aluminum sheet, plate, and foil
  ___ 3354 Aluminum extruded products
  ___ 3355 Aluminum rolling and drawing, n.e.c.*
  ___ 3356 Rolling, drawing, and extruding of nonferrous metals, except copper and aluminum
  ___ 3357 Drawing and insulating of nonferrous wire
  ___ 3363 Aluminum die-castings
  ___ 3364 Nonferrous die-castings, except aluminum
  ___ 3365 Aluminum foundries
  ___ 3366 Copper foundries
  ___ 3369 Nonferrous foundries, except aluminum and copper
  ___ 3398 Metal heat treating
  ___ 3399 Primary metal products, n.e.c.*
___ 34 Fabricated metal products, except machinery and transportation equipment
___ 35 Industrial and commercial machinery and computer equipment

___ 36 Electronic and other electrical equipment and components, except computer equipment

___ 37 Transportation equipment

___ 38 Measuring, analyzing, and controlling instruments; photographic, medical and optical goods; watches and clocks

___ 39 Miscellaneous manufacturing industries

* Not Elsewhere Classified

NOTES

NOTES

PART 2

Environmental Management

NOTES

___ A. Environmental Liability and Hazardous Waste Cleanup

___ 1. Phase 1 Environmental Assessment Protocol checklist (Ref. 12)

    ___ *a.* Project information

        ___ Project property name, address, county/state, and location description

        ___ Property owner name, address, phone

        ___ Key site manager name and phone

        ___ Identification of involved parties, such as lenders, brokers, tenants, and attorneys, including names, addresses, and phone numbers

        ___ Property description, lot size, building(s) size, building(s) age, legal description

        ___ Past property uses and name and address of prior property owner

        ___ Adjoining properties to north, northeast, east, southeast, south, southwest, west, and northwest

        ___ Known contamination on properties in the vicinity of the site

    ___ *b.* Records review

        ___ Review federal environmental record sources, including the following data bases—NPL, CERCLIS, RCRA TSD, RCRA Generators, and ERNS.

        ___ Review state environmental record sources, including state lists of hazardous waste sites, state landfill and/or solid waste disposal site lists, state leaking UST lists, state registered UST lists (property and adjoining properties).

        ___ Review local record sources, such as department of health/environmental division, fire department, planning/zoning department, building permit/inspection department, local/regional pollution control agency, local/regional water quality control agency, local electric utility companies (with respect to PCBs); include contact person's name and phone number.

___ Review local lists, such as lists of landfill/solid waste disposal sites, hazardous waste/contaminated sites, registered underground storage tanks, records of emergency release reports, and records of contaminated public wells.

___ Review standard historical sources, such as aerial photographs, fire insurance maps, property tax files, recorded land title records (mandatory), and national wild life maps for designation of wetlands.

___ Review physical setting sources, such as topographic map (mandatory), hydrology, and hydrogeology maps.

___ Interview persons familiar with the property, such as key site manager and major occupant of property.

___ *c.* Site reconnaissance

___ General information about activities at site, such as raw materials, processes, products, and services; identification of any property contamination and any major environmental concerns, including areas of regulatory uncertainty; identification of environmental incidents or problems on surrounding properties; identification of community complaints

___ Air issues such as air permits, emissions inventory, fugitive emissions, and radon

___ Existence of friable and nonfriable asbestos and lead-based paint

___ Existence of polychlorinated biphenyls (PCBs) in transformers, capacitors, or other potential PCB-containing equipment located on the property; identification of leaking equipment; identification of equipment using hydraulic fluid, such as elevators and hydraulic lifts; evidence of PCB contamination

___ Existence (past and present) of aboveground and underground storage tanks (USTs), including tank capacity, substance stored, installation date, registration, corrosion protection, leak detection, tank construction material, results of testing, date operation terminated, tank removal contractor, visible evidence or records of leaks or failures

___ Information pertaining to hazardous materials handling, including chemicals handled on site, location, manufacturer, and quantities; condition of containers; description of storage areas

___ Information pertaining to hazardous waste storage, treatment, and disposal, including EPA identification numbers, processes that generate hazardous waste, waste types and quantities; facility RCRA status; storage information, including number and size of drums

___ Operations pertaining to treatment, disposal, reclamation, and recycling; permit numbers; transporters and storage and disposal facilities currently used for hazardous waste; inactive land disposal sites owned or operated by the facility; identification of waste disposal sites for which the facility has been identified as a potential contributor; identification of any spill cleanup/soil remediation projects

___ Condition of hazardous waste containers; description of storage area for hazardous waste; identification of hazardous waste disposed with solid waste; identification of visible evidence of hazardous waste disposal problems

___ Description of biological or medical waste generated on property

___ Description of solid waste–handling operations, including solid waste collection and disposal methods

___ Description of waste oil–handling operations, including how waste oil is collected and the ultimate disposition of waste oil, such as recycling, reuse, incineration, and fuel blending; description of container and tank type, capacity, generation rate, and disposal transporter

___ Wastewater treatment issues, including industrial process discharge sources, approximate flow rates, type and number of discharge points, and permits and issuing agency for each discharge; description of pretreatment operations associated with industrial process discharges

___ Stormwater discharges, including approximate volume, area drained, number of discharge points and permits

___ Description of groundwater-monitoring programs on site or near the site, including number and location of wells, latest sample date, frequency of sampling; description of remediation programs on site or near the site

___ Description of sanitary discharges and on-site treatment; identification of septic tanks both operating and abandoned, including location and size

___ Identification of oil/water separators connected to the sanitary system, including clean-out records

___ Description of water wells on site for drinking, process, or cooling—both operating and abandoned—including location and date last used; identification of water source for site; last test of potable water source

___ Description of surface water, standing water, saturated surfaces, ditches, and drainage depressions located on site; flood plain information; description of direction of runoff and runon

___ Issues concerning adjoining properties, including high risk operations such as service stations, dry cleaners, and chemical manufacturing facilities; potential for groundwater contamination from adjoining properties; potential for contaminated stormwater runon from adjoining properties

___ Inclusion of information about photographs, including date, time, location and description of photo, direction faced when taking photo

___ Professional's comments, conclusions, and recommendations

___ d. Costs of Phase I assessments

___ $5000 to $10,000 for small- to medium-sized facilities

___ $10,000 to $25,000 for large facilities, depending on facility location and industrial history

___ 2. SUPERFUND

___ a. Registered SUPERFUND sites listed in 40 CFR Part 300

___ Over 1000 general facilities

___ States with approximately 100 general facilities: New Jersey, New York, and Pennsylvania

___ Over 150 federal facilities

    ___ States with approximately 15 or more federal facilities: California and Washington

___ *b.* Emergency planning and response

    ___ (1) Elements of an emergency plan

        ___ Planning elements such as identification and description of facilities that possess hazardous substances

        ___ Description of operations and designation of persons responsible to implement plan

        ___ Emergency notification procedures, including facility coordinators, community emergency personnel, and government agency personnel

        ___ Methods and procedures to be followed in the event of a release

        ___ Procedures for communication methods among responders

        ___ Procedures detailing equipment and available human resources

        ___ Information on major types of emergency medical services in the area

        ___ Methods of public protection

        ___ Evacuation procedures

        ___ Description of training programs

        ___ Procedures for testing and updating the plan

    ___ (2) Operational response phases for oil removal

        ___ Phase I—Discovery or notification

        ___ Phase II—Preliminary assessment and initiation of action

        ___ Phase III—Containment, countermeasures, cleanup, and disposal

        ___ Phase IV—Documentation and cost recovery

___ 3. Costs of cleanup
    ___ *a.* Cost of sorbents and cleanup kits (Ref. 19)
        ___ (1) Maintenance sorbents
            ___ Pads and rolls: $60 to $90 per set of pads or one roll (absorbs approx. 30 gallons)
            ___ Minibooms: $1.15 to $1.35 per linear foot (absorbs approx. ¼ gallon per linear foot)
            ___ Spill pillows: $2.65 to $3.10 per pillow (absorbs approx. ½ gallon per pillow)
        ___ (2) Chemical sorbents
            ___ Pad and rolls: $80 to $90 per set of pads or one roll (absorbs approx. 30 gallons)
            ___ Minibooms: $2.95 to $3.45 per linear foot (absorbs approx. ¼ gallon per linear foot)
            ___ Spill pillows: $7.00 to $8.30 per pillow (absorbs approx. ½ gallon per pillow)
        ___ (3) Petroleum sorbents
            ___ Pad and rolls: $41 to $55 per set of pads or one roll (absorbs approx. 30 gallons)
            ___ Minibooms: $1.50 to $1.75 per linear foot (absorbs approx. ¼ gallon per linear foot)
            ___ Spill pillows: $8.80 to $10.30 per pillow (absorbs approx. 3.5 gallons per pillow)
        ___ (4) Spill cleanup kits
            ___ Response kits: $17 to $30
            ___ Additional items for use with spill cleanup kits and response kits, sold separately

___ (5) Tarpalins for runoff control (Ref. 27)
   ___ Reinforced polyethylene, 3 mils to 8.5 mils thick—$0.10 to $0.22/sq yd
   ___ Mylar polyester, 7 mils thick—$1.10/sq yd

___ *b.* Costs of hazardous waste pickup/treatment/disposal (Ref. 27)
   ___ Vacuum truck, pickup of 5000 gallons—$100
   ___ Treatment of 5000 gallons (without transportation)—$500 to $2500
   ___ Rental of backhoe, ½ cu yd capacity—$400/day
   ___ Rental backhoe, 3½ cu yd capacity—$2525/day
   ___ Drum packing equipment, with hepa filter, installed—$15,000
   ___ Drum packing equipment, without hepa filter, installed—$12,000
   ___ Industrial shredder—$15,000 to $80,000
   ___ Industrial compactor, 0.5 cu yd, installed—$5600
   ___ Industrial compactor, 1.0 cu yd, installed—$8100
   ___ Industrial compactor, 2.5 cu yd, installed—$14,500
   ___ Industrial compactor, 5 cu yd, installed—$22,750
   ___ Disposal of filled 55-gal drum—$200
   ___ Disposal of bulk material (nonhazardous)—$150/ton
   ___ Disposal of contaminated soil—$100 to $300/cu yd

NOTES

NOTES

B. Pollution Prevention (P2) and Design for Environment (DfE)
(Refs. 15 and 17)

__ 1. *Implementing a P2 program*

    __ *a.* Get top management support for P2

        __ Demonstrate top management commitment for P2 through policy.

        __ Demonstrate endorsement of the policy by all levels of management.

        __ Communicate policy for P2 to all employees.

    __ *b.* Develop a P2 program

        __ Designate a P2 coordinator.

        __ Develop a P2 team.

        __ Develop a written P2 plan.

        __ Set strategies and goals.

        __ Increase employee awareness and involvement.

        __ Train employees.

    __ *c.* Characterize process

        __ Develop process flow diagrams.

        __ Develop material balance.

        __ Identify process units that generate waste.

    __ *d.* Assess wastes and identify opportunities

        __ Identify waste streams.

        __ Prioritize waste streams to be reduced or eliminated.

        __ Generate reduction/elimination options.

        __ Analyze source reduction and source control options.

    __ *e.* Analyze costs

        __ Determine full cost of waste generation, including costs incurred in producing and handling the waste as well as treatment and disposal costs.

        __ Develop economic matrix to determine economic feasibility of projects.

      ____ Consider benefits, such as reduced long-term liability, reduced worker exposure to toxic chemicals, and improved community relations.

      ____ Establish cost allocation system to assign responsibility to the process or department that generates the waste.

  ____ *f.* Identify and implement P2 options

      ____ Assess technical feasibility.

      ____ Evaluate economics of project.

      ____ Implement projects.

  ____ *g.* Evaluate program

      ____ Assess commitment.

      ____ Assess progress.

  ____ *h.* Sustain program

      ____ Reemphasize economic benefits.

      ____ Rotate members of pollution prevention team.

      ____ Provide refresher training.

      ____ Reward success.

      ____ Publicize success.

____ 2. Best management practices for P2 (Ref. 18)

  ____ *a.* Inventory management

      ____ Use of inventory control for setting source reduction goals

      ____ Use of material control for optimizing recycling options

  ____ *b.* Production process modification

      ____ Optimize operation and maintenance procedures for source reduction.

      ____ Substitute less hazardous materials.

      ____ Modify equipment and processes for more environmentally friendly manufacturing and cleaner technologies.

  ____ *c.* Segregation of waste

      ____ Segregation of hazardous and nonhazardous waste for easy waste classification and recycling potential

___ Segregation of solvents for easy recycling and reclamation

___ Segregation of liquid and solid wastes for avoiding contamination of recyclable waste

___ d. Preventive maintenance program

   ___ Institute a well-documented preventive maintenance program to minimize risk of accidental releases.

   ___ Keep vendor maintenance manuals handy to ensure use of compatible materials and proper repair procedures.

___ e. Training programs

   ___ Provide operational training of line employees to minimize energy, water, and chemical use.

   ___ Provide material compatibility training to minimize the risk of accidental release.

   ___ Provide training for detecting out-of-spec operations that emit material to the air, water, or land.

   ___ Provide emergency procedures and training for minimizing lost materials during accidents/incidents.

___ f. Effective management practices

   ___ Centralize waste management, where practicable, so as to maximize consolidation/segregation options.

   ___ Set and track pollution prevention/waste reduction targets and goals.

   ___ Strive for continual improvement of operations and waste management.

___ g. Employee participation

   ___ Solicit and reward employee suggestions for waste reduction.

   ___ Ensure that every employee who has the potential to affect waste reduction activities understands that his or her contribution toward pollution prevention is important.

_____ *h.* Production planning and scheduling

_____ Plan clean-out activities to minimize waste.

_____ Plan production runs to minimize water, energy, and chemical usage.

_____ Schedule batch operations to minimize waste.

_____ *i.* Audit progress

_____ Review water, energy, and chemical usage on a periodic basis.

_____ Review waste generation/minimization on a periodic basis.

_____ Share results with managers and employees whose efforts affect the pollution prevention program.

_____ 3. *P2 programs and information sources* (Ref. 43)

_____ *a.* American Institute for Pollution Prevention

_____ Sponsored by EPA and DOE

_____ Serves as communication bridge to its members

_____ Promotes policies and provides for transfer of information

_____ Proactively sets future directions through cooperative and collaborative efforts among industry, government, and the public

_____ *b.* Enviroene

_____ Funded by EPA and the Strategic Environmental Research and Development Program

_____ Provides a wide collection of pollution prevention information for 18 industrial sectors

_____ World Wide Web address http://environsense.com; also can be found on EPA bulletin board

___ *c.* National Pollution Prevention Center for Higher Education
  ___ EPA-sponsored; located at the University of Michigan
  ___ Collects, develops, and disseminates educational materials on pollution prevention
  ___ Offers internship program, professional education and training, and conferences

___ *d.* P2Info
  ___ DOE-sponsored; operated by Pacific Northwest laboratory
  ___ Provides a clearinghouse on pollution prevention technologies and vendors

___ *e.* Pacific Northwest Pollution Prevention Research Center
  ___ Nonprofit organization serving EPA Region X (Alaska, Idaho, Oregon, and Washington)
  ___ Publishes bimonthly newsletter called *Pollution Prevention Northwest* that covers industry-specific research and state pollution prevention programs

___ *f.* Pollution Prevention Information Clearinghouse
  ___ EPA-sponsored
  ___ Offers EPA documents and fact sheets about pollution prevention
  ___ Offers Pollution Prevention Directory

___ *g.* Waste Reduction Innovative Technology Evaluation
  ___ EPA-sponsored
  ___ Focuses on pilot projects in pollution prevention performed in cooperation with state and local governments
  ___ Program designed to assist federal, state, and local government, as well as small and midsized industries

___ *h.* Waste Reduction Resource Center

    ___ Focuses on industries in EPA Regions III and IV (Alabama, Delaware, the District of Columbia, Florida, Georgia, Kentucky, Maryland, Mississippi, North Carolina, Pennsylvania, South Carolina, Tennessee, Virginia, and West Virginia)

    ___ Information on pollution prevention, including articles, case studies, and technical reports

___ 4. P2 for household hazardous wastes

    ___ *a.* General considerations

        ___ Use products of less toxicity.

        ___ Before buying a product, estimate the amount needed and buy that size.

        ___ Use the product correctly.

    ___ *b.* Specific considerations

        ___ Use less hazardous cleaners such as baking soda, vinegar, borax detergents, and lemon juice.

        ___ Use water-based or latex paints, recycled paints, or natural earth pigment finishes.

        ___ Reduce need for fungicides by keeping areas clean, dry, and not overwatered.

        ___ Use naturally derived pesticides, such as pyrethrum, rotenone, sabadilla, nicotine, and insecticidal soap.

___ 5. Driving forces influencing adoption of DfE

    ___ International standards

    ___ Competitive pressures

    ___ Enterprise integration

    ___ Sustainable development

    ___ Risk management

    ___ Product stewardship

    ___ Regulatory constraints

    ___ Customer satisfaction

____ 6. DfE as a crossroad between the drive toward enterprise integration and sustainable development (See Fig. 2.B.1.) (Ref. 15)

Figure 2.B.1

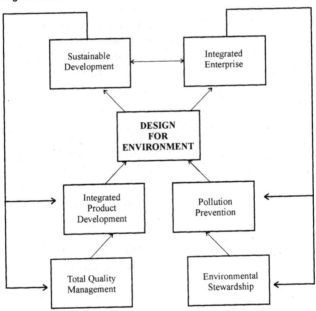

____ 7. DfE voluntary programs
    ____ *a.* Green Lights
        ____ EPA-sponsored
        ____ Membership specifies commitment to install energy-efficient lights at their facilities
    ____ *b.* Energy Star
        ____ EPA-sponsored and designed to encourage energy conservation in electronic devices
        ____ Sets maximum power levels for devices such as personal computers, and requires automatic power-down features

___ c. Climate Wi$e
  ___ EPA-sponsored
  ___ Developed to reduce greenhouse gas emissions
___ d. Waste Wi$e
  ___ EPA-sponsored
  ___ Developed to reduce municipal solid waste
___ e. Environmental Leadership Program
  ___ EPA-sponsored
  ___ Voluntary program that recognizes industries for environmental leadership efforts
___ f. Project XL
  ___ EPA-sponsored
  ___ Voluntary program that recognizes innovative approaches to DfE, pollution prevention, and environmental management
___ g. Common Sense Initiative
  ___ EPA-sponsored initiative with selected industrial sector participants
  ___ Provides voluntary approaches to DfE and pollution prevention
___ 8. DfE guidelines
  ___ a. Design for material and waste recovery
    ___ Avoid composite materials.
    ___ Specify recyclable materials.
    ___ Use recyclable packaging.
    ___ Design for waste recovery and reuse.
  ___ b. Design for component recovery
    ___ Design reusable containers.
    ___ Design for refurbishment.
    ___ Design for remanufacture.
  ___ c. Design for disassembly
    ___ Optimize disassembly sequence.
    ___ Design for easy removal.
    ___ Avoid embedded parts.
    ___ Simplify interfaces.
    ___ Reduce product complexity.

___ Reduce the number of parts.
___ Design for multifunctional parts.
___ Utilize common parts.
___ d. Design for source reduction
___ Reduce product dimensions.
___ Specify lighter-weight materials.
___ Design thinner enclosures.
___ Reduce mass of components.
___ Reduce packaging weight.
___ Use electronic documentation.
___ e. Design for separability
___ Facilitate identification of materials.
___ Use fewer types of materials.
___ Use similar or compatible materials.
___ f. Design for energy conservation
___ Reduce energy use in production.
___ Reduce device power consumption.
___ Reduce energy use in distribution.
___ Use renewable forms of energy.
___ g. Design for material conservation
___ Design multifunctional products.
___ Specify recycled or renewable materials.
___ Use remanufactured components.
___ Design for product longevity.
___ Design for closed-loop recycling.
___ Design for packaging recovery.
___ h. Design for chronic risk reduction
___ Reduce production releases to the environment.
___ Avoid hazardous/toxic substances.
___ Avoid ozone-depleting chemicals.
___ Use water-based technologies.
___ Assure product biodegradability.
___ Assure waste disposability.
___ i. Design for accident prevention
___ Avoid caustic and/or flammable materials.
___ Provide pressure relief mechanisms.
___ Minimize leakage potential.

NOTES

__ C. Storage and Containment of Hazardous Materials and Hazardous Waste

__ 1. Aboveground storage tank systems
    __ *a.* Design concepts
        __ Maximum inspectability
        __ Greater than 100% capacity for secondary containment
        __ Leak detection
        __ Elimination of penetrations through secondary containment
        __ Containment of pipes and pumps
    __ *b.* Costs of aboveground tank systems (not including pipe, pumps, or other equipment) (Ref. 27)
        __ 250,000-gal tank—$310,000
        __ 500,000-gal tank—$380,000
        __ 1,000,000-gal tank—$510,000
    __ *c.* Selected tank-testing methods published by American Society for Testing and Materials (ASTM)
        __ ASTM A 275/A 275M-94—*Method for Magnetic Particle Examination of Steel Forgings*
        __ ASTM A 754-79(1990)—*Test Method for Coating Thickness by X-ray Fluorescence*
        __ ASTM C 868-85(1990)—*Standard Test Method for Chemical Resistance of Protective Linings*
        __ ASTM C 982-88(1992)—*Guide for Selecting Components for Generic Energy Dispersive X-Ray Fluorescence Systems for Nuclear Related Material Analysis*
        __ ASTM C 1118-89(1994)—*Guide for Selecting Components for Wavelength-Dispersive X-Ray Fluorescence Systems*
        __ ASTM D 471-95—*Test Method for Rubber Property–Effect of Liquids*

—— ASTM D 3491-92—*Standard Methods of Testing Vulcanizable Rubber Tank and Pipe Lining*

—— ASTM E 125-63(1993)—*Reference Photographs for Magnetic Particle Indications on Ferrous Castings*

—— ASTM E 165-95—*Practice for Liquid Penetrant Examination*

—— ASTM E 432-91—*Guide for the Selection of a Leak Testing Method*

—— ASTM E 433-71(1993)—*Reference Photographs for Liquid Penetrant Inspection*

—— ASTM E 709-95—*Guide for Magnetic Particle Examination*

—— ASTM E 750-88—*Practice for Characterizing of Acoustic Emission Instrumentation*

—— ASTM E 1002-94—*Method for Testing for Leaks Using Ultrasonics*

—— ASTM E 1003-84(1994)—*Method for Hydrostatic Leak Testing*

—— ASTM E 1065-92—*Guide for Evaluating Characteristics of Ultrasonic Search Units*

—— ASTM E 1067-89—*Practice for Acoustic Emission Testing of Fiberglass-Reinforced Plastic Resin Tanks / Vessels*

—— ASTM E 1139-92—*Practice for Continuous Monitoring of Acoustic Emission from Metal Pressure Boundaries*

—— ASTM E 1172-87(1992)—*Practice for Describing and Specifying a Wavelength-Dispersive X-Ray Spectrometer*

—— ASTM E 1208-94—*Test Method for Fluorescent Liquid Penetrant Examination Using the Lipophilic Post-Emulsification Process*

___ ASTM E 1209-94—*Test Method for Fluorescent Penetrant Examination Using the Water-Washable Process*

___ ASTM E 1210-94—*Test Method for Fluorescent Penetrant Examination Using the Hydrophilic Post-Emulsification Process*

___ ASTM E 1211-87(1992)—*Practice for Leak Detection and Location Using Surface-Mounted Acoustic Emission Sensors*

___ ASTM E 1219-94—*Test Method for Fluorescent Penetrant Examination Using the Solvent-Removable Process*

___ ASTM E 1220-92—*Test Method for Visible Penetrant Examination Using the Solvent-Removable Process*

___ ASTM G 1-90(1994)—*Recommended Practice for Preparing, Cleaning, and Evaluating Corrosion Test Specimens*

___ ASTM G 4-95—*Guide for Conducting Corrosion Coupon Tests in Plant Equipment*

___ ASTM G 5-94—*Reference Test Method for Making Potentiostatic and Potentiodynamic Anodic Polarization Measurements*

___ ASTM G 15-93—*Terminology Relating to Corrosion and Corrosion Testing*

___ ASTM G 41-90(1994)—*Practice for Determining Cracking Susceptibility of Metals Exposed Under Stress to a Hot Salt Environment*

___ ASTM G 46-94—*Recommended Practice for Examination and Evaluation of Pitting Corrosion*

___ ASTM G 48-92—*Testing for Pitting and Crevice Corrosion Resistance of Stainless Steels and Related Alloys by Use of Ferric Chloride*

___ ASTM G 49-85(1990)—*Recommended Practice for Preparation and Use of Direct Tension Stress Corrosion Test Specimen*

___ ASTM G 50-76(1992)—*Recommended Practice for Conducting Atmospheric Corrosion Tests on Metals*

___ ASTM G 51-95—*Test Method for pH of Soil for Use in Corrosion Testing*

___ ASTM G 62-87(1992)—*Test Method for Holiday Detection in Pipeline Coatings*

___ ASTM G 71-81(1992)—*Practice for Conducting and Evaluating Galvanic Corrosion Tests in Electrolytes*

___ ASTM G 78-95—*Guide for Crevice Corrosion Testing of Iron Base and Nickel Base Stainless Alloys in Seawater and Other Chloride-Containing Aqueous Environments*

___ ASTM G 82-83(1993)—*Guide for Development and Use of a Galvanic Series for Predicting Corrosion Performance*

___ ASTM G 90-94—*Practice for Performing Accelerated Outdoor Weathering of Nonmetallic Materials Using Concentrated Natural Sunlight*

___ ASTM G 97-89—*Test Method for Laboratory Evaluation of Magnesium Sacrificial Anode Test Specimens for Underground Applications*

___ ASTM G 96-90—*Guide for On-Line Monitoring of Corrosion in Plant Equipment* (Electrical and Electrochemical Methods)

___ ASTM G 101-94—*Guide for Estimating the Atmospheric Corrosion Resistance of Low-Alloy Steels*

___ ASTM G 104-89(1993)—*Test Method for Assessing Galvanic Corrosion Caused by the Atmosphere*

___ *d.* Documents related to tank systems published by American Petroleum Institute (API)

    ___ API Recommended Practice 2X (1988)—*Recommended Practice for Ultrasonic Examination of Offshore Structural Fabrication and Guidelines for Qualification of Ultrasonic Technicians,* 2nd ed. (ANSI/API RP 2X–1992)

    ___ API Publication 301 (1989,1991)—*Aboveground Storage Tank Survey*

    ___ API Publication 306 (1991)—*An Engineering Assessment of Volumetric Methods of Leak Detection in Aboveground Storage Tanks*

    ___ API Publication 307 (1991)—*An Engineering Assessment of Acoustic Methods of Leak Detection in Aboveground Tanks*

    ___ API Standard 510 (1992)—*Pressure Vessel Inspection Code: Maintenance Inspection, Rating, Repair, and Alteration,* 7th ed. (ANSI/API 510–1992)

    ___ API Recommended Practice 520 (1993)—*Sizing, Selection, and Installation of Pressure Relieving Systems in Refineries, Part 1,* "Sizing and Selection," 6th ed. (ANSI/API Std 520&520-1–1992)

    ___ API Recommended Practice 520 (1994)—*Sizing, Selection, and Installation of Pressure Relieving Systems in Refineries, Part 2,* "Installation," 4th ed.

    ___ API Recommended Practice 521 (1990)—*Guide for Pressure-Relieving and Depressuring Systems,* 3d ed. (ANSI/API RP 521–1992)

    ___ API Standard 620 (1990)—*Design and Construction of Large, Welded, Low-Pressure Storage Tanks,* 8th ed. (ANSI/API Std 620–1992)

— API Standard 650 (1993)—*Welded Steel Tanks for Oil Storage,* 9th ed. (ANSI/API Std 650–1992)

— API Recommended Practice 651 (1991)— *Cathodic Protection of Aboveground Petroleum Storage Tanks*

— API Recommended Practice 652 (1991)— *Lining of Aboveground Petroleum Storage Tank Bottoms* (ANSI/API Std 652–1992)

— API Standard 653 (1991)—*Tank Inspection, Repair, Alteration, and Reconstruction* (Supplement 1, January 1992) (ANSI/API Std 653–1992)

— API Recommended Practice 1110 (1991)— *Recommended Practice for the Pressure Testing of Liquid Petroleum Pipelines,* 3d ed.

— API Recommended Practice 1621 (1993)—*Recommended Practice for Bulk Liquid Stock Control at Retail Outlets,* 5th ed.

— API Recommended Practice 1631 (1992)—*Interior Lining of Underground Storage Tanks,* 3d ed.

— API Publication 2015 (1994)—*Safe Entry and Cleaning of Petroleum Storage Tanks, Planning and Managing Tank Entry from Decommissioning Through Recommissioning,* 5th ed. (ANSI/API Std 2015–1994)

— API Recommended Practice 2350 (1987)— *Overfill Protection for Petroleum Storage Tanks*

— 2. Underground storage tank (UST) systems

   — a. Requirements for USTs containing regulated materials (RCRA)

      — Tanks must be properly designed and constructed, and any portion underground that routinely contains product must be protected from corrosion.

___ Piping that routinely contains regulated substances and is in contact with the ground must be properly designed, constructed, and protected from corrosion.

___ Spill and overfill prevention equipment associated with product transfer to the UST must be designed to prevent a release of the product to the environment during transfers (i.e., use of catch basins, automatic shut-offs, flow restriction devices, alarms, and other equipment).

___ Tanks and piping must be properly installed in accordance with a nationally recognized code of practice.

___ The installation must be performed by a certified installer and completed properly.

___ *b.* Design concepts

   ___ (1) Secondary containment (See Figs. 2.C.1 and 2.C.2.) (Ref. 42)

      ___ Design should provide an outer barrier between the tank and the soil and backfill capable of holding the material long enough so that a release can be detected.

Figure 2.C.1

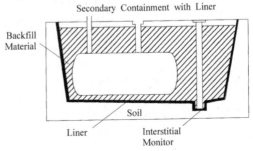

Secondary Containment with Liner

Backfill Material

Soil

Liner

Interstitial Monitor

___ Design should prevent migration of contaminant.

___ Types include double-walled tank, jacketed tank, and tank with fully enclosed external liners.

Figure 2.C.2

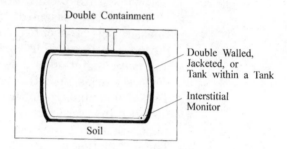

Double Containment

Double Walled, Jacketed, or Tank within a Tank

Interstitial Monitor

Soil

___ (2) Interstitial monitoring

    ___ System should be designed to detect any leak from the underground tank under normal operating conditions.

    ___ Leak detection mechanisms include electrical conductivity, pressure sensing, fluid sensing, hydrostatic monitoring, manual inspection, and vapor monitoring.

___ (3) Other release detection methods

    ___ Inventory control

    ___ Vapor monitoring

    ___ Groundwater monitoring

    ___ Tank gauging

    ___ Tank tightness testing using volumetric tests

___ c. Costs of underground storage tank systems, fiberglass, double-walled (not including piping, pumps, and other equipment) (Ref. 27)

    ___ 600 gal—$3200

___ 1000 gal—$4300

___ 2500 gal—$6250

___ *d.* Costs of removal of underground storage tanks (includes excavation and loading onto trailer) (Ref. 27)

   ___ Nonleaking tank, 3000 to 5000–gal capacity—$300 to $450

   ___ Leaking tank, 3000 to 5000–gal capacity—$600 to $900

   ___ Nonleaking tank, 6000 to 8000–gal capacity—$400 to $650

   ___ Leaking tank, 6000 to 8000–gal capacity—$800 to $1,300

   ___ Nonleaking tank, 9000 to 12,000–gal capacity—$625 to $950

   ___ Leaking, 9000 to 12,000–gal capacity—$1250 to $1900

   ___ Haul tank to salvage dump, 100 miles round-trip—$500 to $900

___ *e.* Standards pertaining to USTs

   ___ API Publication 1632 (1987)—*Cathodic Protection of Underground Petroleum Storage Tanks and Piping Systems,* 2nd ed.

   ___ API Recommended Practice 1615 (1987)—*Installation of Underground Petroleum Storage Systems,* 4th ed.

   ___ ASTM D 4021-92—*Specification for Glass-Fiber-Reinforced Polyester Underground Petroleum Storage Tanks*

   ___ ASTM E 1430-91—*Guide for Use of Release Detection Devices with Underground Storage Tanks*

   ___ ASTM E 1526-93—*Practice for Evaluating the Performance of Release Detection Systems for Underground Storage Tank Systems*

___ 3. Container storage
    ___ *a.* Hazardous waste
        ___ (1) *Definition of "empty" containers (RCRA)*
            ___ All material has been removed, as reasonably expected with the type of material.
            ___ There is less than one inch of residue on the bottom, or
            ___ If the container volume is 110 gallons or less, the residue is no more than 3% by weight of the total container capacity, or
            ___ If the container volume is more than 110 gallons, the residue is no more than 0.3% by weight of the total container capacity.
            ___ For compressed gas containers, the inside pressure must be nearly atmospheric.
        ___ (2) *Definition of "empty" for container holding acutely hazardous waste (RCRA)*
            ___ The container has been triple-rinsed with a solvent capable of removing the material.
            ___ Another equivalent method has been used.
            ___ The liner of the container has been removed so that the outer container is considered empty.
    ___ *b.* Design concepts for container storage areas
        ___ Impervious base or floor coating
        ___ Adequate floor drainage and mechanisms to collect leaks, spills, and precipitation, or elevation of containers to disallow contact with leaks, spills, and precipitation
        ___ Secondary containment large enough to hold 10% of the total volume or the vol-

ume of the largest container, whichever is greater

___ Design which prevents the entry of stormwater

___ c. Good management practices

___ Segregate incompatible waste containers.

___ Wash containers before reuse.

___ Keep container closed at all times, except when adding or removing waste.

___ Inspect containers for damage, corrosion, or other signs of deterioration.

___ Control air emissions per regulations (40 CFR 264, Subpart CC).

___ When storage area is no longer in use, close the facility properly.

___ 4. Tank and container material compatibility (Refs. 35 and 42)

___ a. Acids

___ (1) Compatible with

___ Austenitic stainless steels such as AISI 316 and 317

___ Several nickel-based alloys with molybdenum

___ Platinum alloys

___ Tantalum, titanium, and hastelloy

___ Polyester-fiberglass

___ Glass-lined steel

___ (2) Incompatible with

___ Cast iron

___ Carbon steel

___ Other iron-based metals

___ b. Alcohols

___ (1) Compatible with

___ Carbon steel

___ Stainless steel

___ Cast iron

___ Aluminum

___ (2) Incompatible with
    ___ Many rubbers
    ___ Many plastics
___ c. *Aldehydes*
    ___ (1) Compatible with
        ___ Stainless steels
        ___ Nickel-based alloys
        ___ Copper-based alloys
    ___ (2) Incompatible with
        ___ Plastics
        ___ Most rubbers
___ d. *Ammonium solutions*
    ___ (1) Compatible with
        ___ Carbon steel
        ___ Stainless steel
        ___ Nickel-based alloys
        ___ Synthetic and natural rubber
        ___ Some plastics
    ___ (2) Incompatible with copper-based metals
___ e. *Caustics*
    ___ (1) Compatible with
        ___ Carbon steel
        ___ Stainless steel
        ___ Nickel and nickel-based alloys
        ___ Natural and butyl rubbers
        ___ Fiberglass
    ___ (2) Incompatible with
        ___ Aluminum
        ___ Some iron-based metals
___ f. *Petroleum distillates*
    ___ (1) Compatible with
        ___ Carbon steel
        ___ Other metals
    ___ (2) Incompatible with
        ___ Synthetic and natural rubbers
        ___ Polyethylene
        ___ Polypropylene

NOTES

NOTES

D. Waste Treatment and Disposal Technologies (Refs. 38, 42, and 43)

___ 1. *Physical treatment*
   ___ *a. Stripping*
      ___ (1) Air sparging
         ___ This process is used to remove insoluble organics from water.
         ___ Air is forced through tank to create bubbles; the volatile organics are stripped from the liquid phase and are transferred to the gas phase.
      ___ (2) Countercurrent stripping (See Fig. 2.D.1.) (Ref. 42)
         ___ Air stripping is accomplished in a column.
         ___ The column height and diameter are sized to provide sufficient contact time for adequate contaminant removal.

Figure 2.D.1

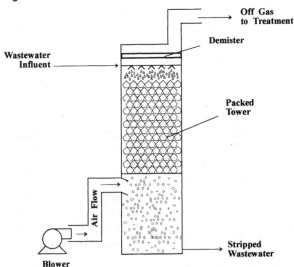

___ (3) Steam stripping (See Fig. 2.D.2.)

___ Used to strip more concentrated volatile organic compounds from aqueous wastewaters

___ Similar to continuous fractional distillation

Figure 2.D.2

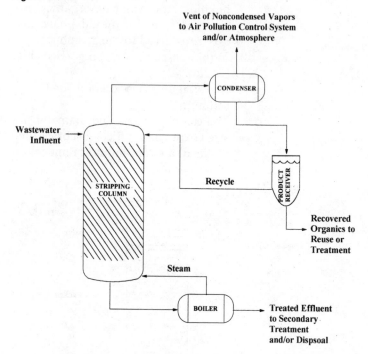

___ *b. Distillation*

___ (1) Batch distillation (See Fig. 2.D.3.)

___ Process is used for small lots of solvents that have varied compositions and vapor pressures.

_____ Process uses a heated evaporation chamber or boiler to vaporize the product, which is then withdrawn and condensed for reuse.

Figure 2.D.3

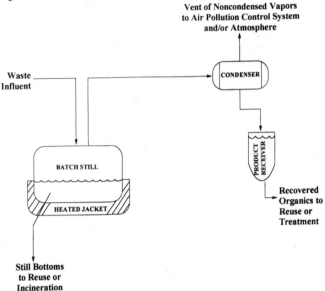

_____ (2) Continuous fractional distillation (See Fig. 2.D.4.)
_____ Used for product contaminated with a material of similar vapor pressure
_____ Accomplished in large tray columns or packed towers, with several stages of distillation
_____ Cannot be used to treat liquids with high viscosity at high temperatures or liquids with a high solids concentration

Figure 2.D.4

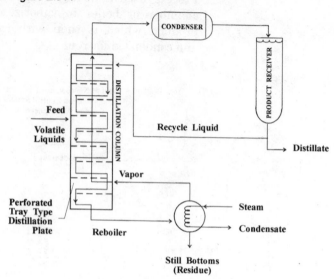

___ c. *Evaporation*

    ___ (1) Thin-film evaporators

        ___ Use centrifugal force or rotating wiper blades to spread film on heated wall.

        ___ Volatile compounds are driven off and collected on an internal chilled condenser surface.

    ___ (2) Dryers

        ___ Used for waste streams of very high-solids content

        ___ Types include vacuum rotary dryer, drum dryer, tray and compartment dryer, and continuously operating conveying dryer (See Fig. 2.D.5.)

    ___ (3) Aqueous evaporation

        ___ Process is used in treatment of wastewaters containing salts and dissolved solids.

___ Evaporation minimizes the volume
of waste requiring disposal.

Figure 2.D.5

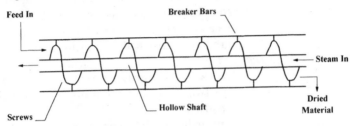

___ d. *Membrane filtration*
  ___ Includes a number of well-known
  processes, including reverse osmosis,
  ultrafiltration, nanofiltration, microfil-
  tration, and electrodialysis
  ___ Uses a driving force (electricity or pressure) to filter particles, ions, and organic
  molecules through a membrane, producing a clean stream on one side and a concentrated stream on the other
___ e. *Sedimentation/clarification* (See Fig. 2.D.6.)

Figure 2.D.6

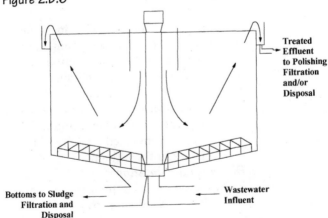

—— Process uses gravity settling to remove hazardous and nonhazardous grits, fines, and other suspended solids.

—— Heavier solids settle to bottom and are removed and thickened prior to disposal.

—— Flocculating agents may be used in clarifiers to enhance sedimentation.

—— f. *Decantation*

—— Process uses a gravity decanter to separate two immiscible liquids of different densities.

—— Lighter liquid floats on top and is drawn off through an overflow drain; heavier liquid is drawn off through a bottom drain.

—— g. *Carbon adsorption*

—— Process uses activated carbon to remove dilute organics from a waste stream.

—— Carbon beds can be upflow or downflow columns or can be open beds.

—— Breakthrough of waste occurs when beds are saturated (See Fig. 2.D.7).

—— Carbon beds typically are applied in series.

—— h. *Ion exchange*

—— An adsorption process in which ionic species are adsorbed from solution by changing places with similarly charged ions on the exchange media

—— Primarily used to remove metals

—— Can be operated in parallel or series and in downflow or upflow (fluidized) mode

—— Beds regenerated when target ion appears in the effluent

—— i. *Dissolved air flotation*

—— Process is used to separate suspended solids and oil and grease from aqueous streams and to concentrate or thicken sludges.

___ Process is especially effective for emulsions after they have been chemically broken.

___ Process uses air bubbles to carry or float the solids and oils to the surface, from which they can be removed.

Figure 2.D.7

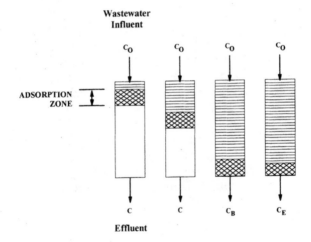

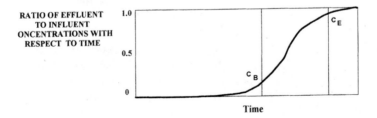

____ *j. Extraction*
____ (1) Solvent extraction
____ Process involves mixing of a waste stream, which contains a hazardous constituent (solute) to be extracted, with an extraction fluid (solvent); extraction occurs when the solute has greater solubility in the extraction fluid than in the waste stream.
____ The waste stream and solute-enriched solvent (extract) form distinct phases, and a continuous gravity decanter can be used to separate and decant the extract.
____ (2) Supercritical fluid extraction
____ Process uses pressurized or supercritical gas such as carbon dioxide as the solvent.
____ Efficiency of the extraction process is increased through improved mass transfer rates.
____ *k. Stabilization/solidification*
____ (1) Cement, lime, and pozzolanic materials
____ Materials form a calcium aluminosilicate crystalline structure that can physically and/or chemically bind with toxic constituents such as metal hydroxides and carbonates.
____ The high pH of the cementitious binders protects against acid conditions that can cause metal leaching.
____ (2) Thermoplastics
____ Used for waste containing heavy metals and/or radioactive materials
____ Involves mixing waste with a molten thermoplastic material such as bitumen, asphalt, polyethylene, or polypropylene

___ Not suitable for organic or hygroscopic wastes; not suitable for strongly oxidizing constituents, anhydrous inorganic salts, or aluminum salts

___ (3) Thermosetting reactive polymers

___ Materials used include reactive monomers such as urea-formaldehyde, phenolics, polyesters, epoxides, and vinyls that form a polymerized material when mixed with a catalyst.

___ Treatment forms a spongelike material that traps solid particles.

___ Treatment is used for radioactive materials and acid wastes.

___ (4) Polymerization

___ Used for spills of chemicals that are monomers or low-order polymers

___ Involves the addition of a catalyst

___ Used to treat aromatics, aliphatics, and other oxygenated monomers such as vinyl chloride and acrylonitrile

___ 2. *Chemical treatment*

___ *a. Neutralization*

___ Process is used to treat waste acids and waste caustics (bases) in order to eliminate their reactivity and/or corrosiveness.

___ An alkali (caustic) is used to treat acids, and acids are used to treat caustics.

___ Process is performed in a well-mixed tank.

___ *b. Chemical oxidation/reduction*

___ (1) Chlorination

___ Process is used to treat organic compounds, alcohols, and phenols.

___ Addition of chlorine to wastewaters allows chlorine to attach to double carbon bond chemicals to render them less toxic.

___ Chlorination can also be accomplished with the addition of chlorine dioxide, which is a less hazardous substance than chlorine.

___ (2) Ozonation

    ___ Process involves injecting a wastewater stream with ozone through sparging or bubbling.

    ___ Process provides an effective oxidation reaction for treating organic-containing wastewaters.

    ___ Ozone not consumed in the oxidation reaction decomposes to oxygen.

    ___ High electrical costs are associated with ozone production.

___ (3) Electrolytic recovery

    ___ Process is used to reduce and remove metal ions from solution through electrolytic processing in an electrolytic cell.

    ___ Process is limited to solutions containing cadmium, chromium, copper, lead, tin, zinc, gold, and silver—which are high in the electromagnetic series and therefore readily reducible for deposit on a cathode.

___ c. *Wet air oxidation* (See Fig. 2.D.8.)

    ___ Process involves injecting air or oxygen into a heated and pressurized system to oxidize organic contaminants and some inorganics such as cyanide.

    ___ Process is used for wastewaters and wastes containing greater than 85% water, which are too dilute to incinerate and which contain toxics that cannot be treated biologically.

_____ Process is typically operated at temperatures ranging from 175 to 325°C and pressures ranging from 300 to 3000 psi (lb/in^2).

_____ Applications include waste streams containing hydrocarbons, nitrogen compounds, halogens, sulfur, phosphorus compounds, metals, and cyanide.

Figure 2.D.8

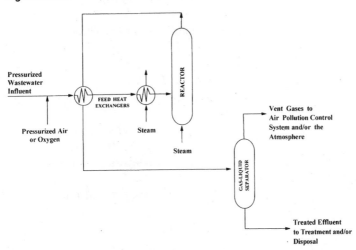

_____ d. *Chemical precipitation*

_____ Process removes metals through an oxidation reaction, which converts the metal to a relatively insoluble metal oxide, hydroxide, or sulfide.

_____ The insoluble metal precipitates out of the wastewater stream as a sludge or sediment that can be dewatered in a centrifuge or filter press and collected for reclamation or disposal.

___ 3. *Biological treatment*

    ___ *a. Factors that influence the biological treatment process*

        ___ Concentration of organics

        ___ Amount of nutrients in the system

        ___ Temperature

        ___ pH

        ___ Treatability of toxic inorganics

        ___ Biodegradability of organics

        ___ Oxygen concentration

        ___ Hydraulic retention time

        ___ Solids reaction time

        ___ Shock loadings of biochemical oxygen demand (BOD), suspended solids, and toxics

    ___ *b. Aerobic biological treatment processes*

        ___ (1) Activated sludge process (See Fig. 2.D.9.)

            ___ Aerobic-suspended growth process widely used for treatment of industrial wastewaters containing organics

            ___ Preceded by equalization to dampen fluctuations in flow and organic concentrations

Figure 2.D.9

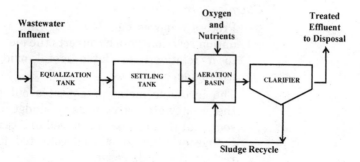

___ (2) Aerobic-supported growth systems
   ___ Rotating biological contactors support a film culture that is attached to a solid surface; treatment occurs as organic-contaminated wastewater is passed over the surface.
   ___ Trickling filter process includes primary clarification, equalization, and treatment through the trickling filters; the packing media of the trickling filters support the microbial film and allow for a large surface area in a small volume.

___ c. Anaerobic processes
   ___ Anaerobic processes break down organics as follows: acid-forming bacteria first convert complex organics into simpler, lower molecular weight—carboxylic acids, alcohols, carbon dioxide, and water; then methane-forming bacteria convert the carboxylic acids to methane gas and carbon dioxide.
   ___ Processes are used for sludges and other more concentrated waste streams.

___ 4. Thermal treatment
   ___ a. Factors affecting performance
      ___ Waste characteristics
      ___ Temperature
      ___ Residence time
      ___ Turbulence
      ___ Air supply
   ___ b. Catalytic oxidation
      ___ Process is used only for gaseous streams, because combustion reactions take place on the surface of the catalyst.
      ___ Common catalysts are palladium and platinum.

——— Systems may be susceptible to poisoning through masking of or interference with the active sites.

——— c. *Fluidized bed*

——— Process involves the waste and an inert bed material that becomes fluidized by blowing heated air through a distributor plat at the bottom of the bed.

——— Process is used to incinerate gases, liquids, or solids, although each type of waste is introduced into the bed differently.

——— Systems are susceptible to seizing or binding up.

——— d. *Liquid injection*

——— Process is commonly used for incineration of liquid hazardous wastes such as solvents.

——— Atomizer breaks liquids into fine droplets, which allows the residence time to be extremely short.

——— Process is not appropriate for high-viscosity liquids.

——— e. *Multiple hearth*

——— Multiple hearths are well-suited for combustion of wet sludges, particularly municipal biological wastewater sludges.

——— Multiple hearths are stacked vertically, and sludges are introduced into the top hearth.

——— A minimum of six hearths are usually required.

——— f. *Rotary kiln* (See Fig. 2.D.10.)

——— System is a long, cylindrical incinerator that is sloped a few degrees from the horizontal; the slope and slow rotation of the kiln continuously mix and reexpose the waste to the hot refractory walls, moving it toward the exit point.

___ System is capable of incinerating a wide range of wastes, including hazardous wastes.

___ g. *Thermal desorption*

___ Process is used primarily for treatment of soils.

___ Heating increases the volatilization of organic contaminants.

Figure 2.D.10

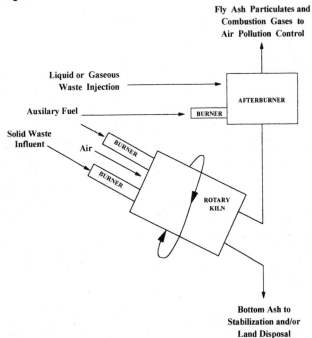

NOTES

___ **E. Wastewater and Stormwater Discharges and Management**

___ 1. *Priority toxic pollutants* (Ref. 4)
    ___ *a. Metals*
        ___ Antimony
        ___ Arsenic
        ___ Beryllium
        ___ Cadmium
        ___ Chromium
        ___ Copper
        ___ Lead
        ___ Mercury
        ___ Nickel
        ___ Selenium
        ___ Thallium
        ___ Zinc
    ___ *b. Volatile organics*
        ___ Acrolein
        ___ Acrylonitrile
        ___ Benzene
        ___ Bromoform
        ___ Carbon tetrachloride
        ___ Chlorobenzene
        ___ Chlorodibromomethane
        ___ Chloroethane
        ___ 2-Chloroethylvinyl ether
        ___ Chloroform
        ___ Dichlorobromomethane
        ___ 1,1-Dichloroethane
        ___ 1,2-Dichloroethane
        ___ 1,1-Dichloroethylene
        ___ 1,2-Dichloropropane
        ___ 1,3-Dichloropropylene
        ___ Ethylbenzene
        ___ Methyl bromide
        ___ Methyl chloride
        ___ Methylene chloride
        ___ 1,1,2,2-Tetrachloroethane

  ___ Tetrachloroethylene
  ___ Toluene
  ___ 1,2-trans-Dichloroethylene
  ___ 1,1,1-Trichloroethane
  ___ 1,1,2-Trichloroethane
  ___ Trichloroethylene
  ___ Vinyl chloride

___ c. *Acid extractables*
  ___ 2-Chlorophenol
  ___ 2,4-Dichlorophenol
  ___ 2,4-Dimethylphenol
  ___ 2-Methyl-4,6-dinitrophenol
  ___ 2,4-Dinitrophenol
  ___ 2-Nitrophenol
  ___ 4-Nitrophenol
  ___ 3-Methyl-4-chlorophenol
  ___ Pentachlorophenol
  ___ Phenol
  ___ 2,4,6-Trichlorophenol

___ d. *Base/neutral extractables*
  ___ Acenaphthene
  ___ Acenaphthylene
  ___ Anthracene
  ___ Benzidine
  ___ Benzo(*a*)anthracene
  ___ Benzo(*a*)pyrene
  ___ Benzo(*b*)fluoranthene
  ___ Benzo(*ghi*)perylene
  ___ Benzo(*k*)fluoranthene
  ___ Bis(2-chloroethoxy)methane
  ___ Bis(2-chloroethyl)ether
  ___ Bis(2-chloroisopropyl)ether
  ___ Bis(2-ethylhexyl)phthalate
  ___ 4-Bromophenyl phenyl ether
  ___ Butylbenzyl phthalate
  ___ 2-Chloronaphthalene
  ___ 4-Chlorophenyl phenyl ether
  ___ Chrysene
  ___ Dibenzo(*a,h*)anthracene

____ 1,2-Dichlorobenzene
____ 1,3-Dichlorobenzene
____ 1,4-Dichlorobenzene
____ 3,3'-Dichlorobenzidene
____ Diethyl phthalate
____ Dimethyl phthalate
____ Di-*n*-butyl phthalate
____ 2,4, Dinitrotoluene
____ 2,6-Dinitrotoluene
____ Di-*n*-octyl phthalate
____ 1,2-Diphenylhydrazine
____ Fluoranthene
____ Fluorene
____ Hexachlorobenzene
____ Hexachlorobutadiene
____ Hexachlorocyclopentadiene
____ Hexachloroethane
____ Indeno(1,2,3-*cd*)pyrene
____ Isophorone
____ Naphthalene
____ Nitrobenzene
____ *n*-Nitrosodimethylamine
____ *n*-Nitrosodi-*n*-propylamine
____ *n*-Nitrosodiphenylamine
____ Phenanthrene
____ Pyrene
____ 1,2,4-Trichlorobenzene
____ *e. Pesticides and PCBs*
____ Aldrin
____ alpha-BHC
____ beta-BHC
____ gamma-BHC
____ delta-BHC
____ Chlordane
____ 4,4'-DDT
____ 4,4'-DDE
____ 4.4'-DDD
____ Dieldrin
____ alpha-Endosulfan

___ beta-Endosulfan
___ Endosulfan sulfate
___ Endrin
___ Endrin aldehyde
___ Heptachlor
___ Heptachlor epoxide
___ PCB-1242
___ PCB-1254
___ PCB-1221
___ PCB-1232
___ PCB-1248
___ PCB-1260
___ PCB-1016
___ Toxaphene
___ *f. Miscellaneous pollutants*
___ Cyanide
___ Asbestos
___ 2,3,7,8-TCDD (dioxin)
___ 2. Industrial categories with federal effluent standards
___ Aluminum forming (40 CFR Part 467)
___ Asbestos manufacturing (40 CFR Part 427)
___ Battery manufacturing (40 CFR Part 461)
___ Builders paper and board mills (40 CFR Part 431)
___ Canned and preserved fruits and vegetables processing (40 CFR Part 407)
___ Metal finishing (40 CFR Part 433)
___ Metal molding and casting (40 CFR Part 464)
___ Mineral mining and processing (40 CFR Part 436)
___ Nonferrous metals (40 CFR Part 421)
___ Offshore oil and gas extraction (40 CFR Part 435)
___ Ore mining and dressing (40 CFR Part 440)
___ Organic chemicals, plastics, and synthetic fibers (40 CFR Part 414)
___ Paint formulating (40 CFR Part 446)
___ Paving and roofing materials (40 CFR Part 443)

___ Pesticide chemicals manufacturing (40 CFR Part 455)

___ Petroleum refining (40 CFR Part 419)

___ Phosphate manufacturing (40 CFR Part 422)

___ Pharmaceutical manufacturing (40 CFR Part 439)

___ Photographic processing (40 CFR Part 459)

___ Plastics molding and forming (40 CFR Part 463)

___ Porcelain enameling (40 CFR Part 466)

___ Pulp, paper, and paper board (40 CFR Part 430)

___ Rubber processing (40 CFR Part 428)

___ Soaps and detergents (40 CFR Part 417)

___ Steam electric power generating (40 CFR Part 423)

___ Sugar processing (40 CFR Part 409)

___ Textile mills (40 CFR Part 410)

___ Canned and preserved seafood processing (40 CFR Part 408)

___ Carbon black manufacturing (40 CFR Part 458)

___ Cement manufacturing (40 CFR Part 411)

___ Coal mining (40 CFR Part 434)

___ Coil coating (40 CFR Part 465)

___ Copper forming (40 CFR Part 468)

___ Dairy products (40 CFR Part 405)

___ Electrical and electronic components (40 CFR Part 469)

___ Electroplating (40 CFR Part 413)

___ Explosives manufacturing (40 CFR Part 457)

___ Feedlots (40 CFR Part 412)

___ Ferroalloy manufacturing (40 CFR Part 424)

___ Fertilizer manufacturing (40 CFR Part 418)

___ Glass manufacturing (40 CFR Part 426)

___ Grain mills (40 CFR Part 406)

___ Gum and wood chemicals manufacturing (40 CFR Part 454)

___ Hospitals (40 CFR Part 460)

___ Ink formulating (40 CFR Part 447)

___ Inorganic chemicals manufacturing (40 CFR Part 415)

___ Iron and steel manufacturing (40 CFR Part 420)
___ Leather tanning and finishing (40 CFR Part 425)
___ Meat products (40 CFR Part 432)
___ Timber products (40 CFR Part 429)

___ 3. Discharge limits/standards for direct dischargers

___ a. *Best Practicable Technology (BPT)*—Technology-based effluent limits for a particular industrial category; defined in 40 CFR 401.16; includes limits for conventional pollutants—such as BOD, TSS, fecal coliform bacteria, pH, and oil and grease—as well as nonconventional pollutants, such as COD and ammonia

___ b. *Best Conventional Technology (BCT)*—Technology-based effluent limits for conventional pollutants, such as BOD and TSS; BCT limits may be more stringent than BPT if higher level of treatment is reasonable

___ c. *Best Available Technology (BAT)*—Technology-based effluent limits for toxic and nonconventional pollutants

___ d. *New Source Performance Standards (NSPS)*—Effluent limits for new direct dischargers; defined in 40 CFR 122.2; typical discharges are those generated by new construction or major modifications to facilities after effluent limits are proposed for a specific industrial category; typically more stringent than BPT, BCT, and BAT

___ 4. Discharge limits/standards for indirect dischargers

___ a. *Pretreatment Standards for Existing Sources (PSES)*—Standards cover all types of pollutants—conventional, nonconventional, and toxic; analogous to BPT and BAT limits for direct dischargers

___ b. *Pretreatment Standards for New Sources (PSNS)*—Standards cover all types of pollutants—conventional, nonconventional, and toxic; analogous to NSPS for direct dischargers

____ 5. Other effluent limits

    ____ *a. Best Management Practices (BMP)*—Procedures that usually address maintenance and good housekeeping to minimize spills and pollutant concentrations in wastewaters

    ____ *b. Best Professional Judgement (BPJ)*—Used when limits cannot be set entirely according to categorical effluent guidelines; also used to set limits for pollutants not covered by effluent guidelines

____ 6. Discharge prohibitions (Ref. 5)

    ____ Discharges that can cause pass-through or interference

    ____ Pollutants that create a fire or explosion hazard in the POTW

    ____ Pollutants that will cause corrosive structural damage to the POTW

    ____ Solid or viscous pollutants in amounts that will cause obstruction in the flow in the POTW

    ____ Any pollutant, including oxygen-demanding pollutants, released in a discharge at a flow rate and/or pollutant concentration that will cause interference with the POTW

    ____ Heat in amounts that will inhibit biological activity in the POTW

    ____ Petroleum oil, nonbiodegradable cutting oil, or products of mineral oil origin in amounts that will cause interference or pass-through

    ____ Pollutants that result in the presence of toxic gases, vapors, or fumes within the POTW in a quantity that may cause acute worker health and safety problems

    ____ Any trucked or hauled pollutants, except at discharge points designated by the POTW

____ 7. Elements of a stormwater pollution prevention plan (Ref. 41)

    ____ *a.* Description of potential stormwater pollution sources

    ___ Identification of the potential sources that may be reasonably expected to add significant amounts of pollutants to stormwater discharges

    ___ Prioritization of risk of sources

___ *b.* Site map

    ___ Outline of drainage area for each stormwater outfall

    ___ Location of existing structural control measures used to reduce pollutants in stormwater runoff

    ___ Location of surface water bodies

    ___ Locations where significant materials are exposed to precipitation

    ___ Locations where major spills or leaks have occurred

    ___ Locations of major activities such as fueling stations, vehicle and equipment maintenance and/or cleaning areas, loading/unloading areas, treatment storage or waste disposal, liquid storage tanks, processing areas, and storage areas

___ *c.* Material inventory

    ___ Identification of materials that have been handled, treated, stored, or disposed of in a manner to allow exposure to stormwater

    ___ Identification of methods and location of on-site storage and/or disposal

    ___ Identification of other exposed materials

    ___ Identification of materials management practices employed to minimize contact of materials and stormwater runoff

    ___ Identification of treatment technology for stormwater

___ *d.* Evaluation of past spills and leaks

    ___ Identification of significant spills and leaks of toxic and/or hazardous pollutants that occurred in areas exposed to precipitation

    ___ Identification of significant spills and leaks of toxic and/or hazardous pollutants that resulted in discharge to a stormwater drain

___ *e.* Identification of nonstormwater discharges and illicit connections

    ___ Identification of nonstormwater discharges through visual inspections, review of sewer maps, and dye testing; discharges from firefighting activities, fire hydrant flushings and irrigation drainage, routine building/pavement washdown that does not involve the use of detergents, and air conditioning condensate are exempted.

    ___ Identification of illicit connections through roof drain drawings, process piping drawings, and sanitary sewer drawings.

___ *f.* Stormwater quality sampling data

    ___ Summarize existing data.

    ___ Plan for additional sampling events as necessary for proper evaluation of stormwater quality.

___ *g.* Best management practices

    ___ Good housekeeping practices

    ___ Preventive maintenance activities

    ___ Visual inspection program

    ___ Spill prevention and response program

    ___ Sediment and erosion control

    ___ Management of runoff

    ___ Employee training

    ___ Recordkeeping and reporting

NOTES

___ F. Groundwater and Soil Monitoring
(Refs. 39 and 40)

___ 1. Steps in designing a groundwater-monitoring system
  ___ Define regulatory requirements and technical objectives.
  ___ Conduct preliminary investigation.
  ___ Develop initial conceptual model to be used as the basis of the field investigation.
  ___ Conduct field investigation.
  ___ Refine conceptual model to be used as the basis of the monitoring system design.
  ___ Design groundwater monitoring system.
  ___ Install groundwater monitoring system.
  ___ Collect, analyze, and evaluate groundwater samples and data.
  ___ Evaluate the groundwater monitoring system with respect to the regulatory requirements and technical objectives.
  ___ Refine the groundwater monitoring system as necessary.

___ 2. Groundwater-monitoring wells installation
  ___ *a.* Drilling/soils-sampling methods
    ___ (1) *Hollow-stem augers*
      ___ Description—Uses "giant screw" and continuous flighting; as augers advance downward, the cutting moves upward; hollow stem or core allows drill rods and samplers to be inserted through the center of the augers.
      ___ Applications—All types of soil investigations, including unconsolidated formations and stable formations; allows for soil sampling with split spoon or thin wall samplers; applications include water quality sampling.

___ Limitations—Preserving sample integrity difficult in unconsolidated formations; possible cross contamination of aquifers where annular space is not positively controlled by water, drilling mud, or surface casing; sand and gravel heaving may be difficult to control; limited diameter of augers limits casing size; smearing of clays may seal off aquifer to be monitored.

___ (2) *Solid-stem augers*

___ Description—Method similar to that of hollow stem augers; auger made of solid steel and therefore needs to be removed from the borehole to collect undisturbed split spoon or thin wall samples and to install casing.

___ Applications—Shallow soils investigations; soil samples; vadose zone– monitoring wells (lysimeters); monitoring wells in saturated, stable soils; identification of depth to bedrock

___ Limitations—Split spoon or thin wall samplers must be used for sample integrity; soil sample data limited to areas and depths where stable soils are predominant; unsuitable for most unconsolidated aquifers because of bore caving upon auger removal; depth capacity decreases as diameter of auger increases; monitoring well diameter limited by auger diameter.

___ (3) *Cable tool drilling*

___ Description—Utilizes a drilling rig and interlocking steel hammers (jars) which slide independently of each other; the hammering action drives of the jars drive the sampling/well drilling barrel into the ground.

___ Applications—Drilling in all types of geologic formations; accommodates almost any depth and diameter range; excellent samples can be obtained from coarse-grained materials.

___ Limitations—Heaving of unconsolidated materials must be controlled; equipment availability limited in some parts of the United States; drilling process is relatively slow.

___ (4) *Air rotary drilling*

___ Description—Involves the use of circulating fluids such as mud, water, or air to remove the drill cutting and to maintain an open hole as drilling progresses; forces air down the drill pipe and back up the borehole to remove the drill cuttings.

___ Applications—Easily   used   for drilling of semiconsolidated and consolidated rock; provides good quality formation samples; allows for easy and quick identification of lithologic changes; allows for identification of water-bearing zones and estimate of water yields.

___ Limitations—Not suitable for unconsolidated formations; air may alter chemical or biological conditions.

___ (5) *Air rotary with casing driver drilling*

___ Description—Uses air rotary drilling technique with the addition of a casing driver.

___ Applications—Commonly used in unconsolidated sands, silts, and clays; also used for drilling in alluvial material; interaquifer cross contamination minimized from casing supports; provides good quality formation samples; minimal formation damage.

___ Limitations—Thin, low-pressure water-bearing zones easily missed; samples may be pulverized; air may modify chemical or biological conditions.

___ (6) *Mud rotary and water drilling*

___ Description—Involves the introduction of drilling fluids (either drilling muds or water) into the borehole through the drill pipe; drilling fluids maintain open hole, provide lubrication to the drill bit, and remove drill cuttings.

___ Applications—Commonly used in clay, silt, and reasonably compacted sand and gravel; allows for split spoon and thin wall sampling in unconsolidated materials; allows for core sampling in consolidated rock; flexible range of tool sizes and depth capabilities.

___ Limitations—Bentonite or other drilling fluid additives may influence quality of groundwater samples; aquifer identification difficult; samples inadequate for monitoring well screen selection; drilling fluid invasion of permeable zones may compromise validity of subsequent monitoring well samples.

___ (7) *Dual-wall reverse circulation*

    ___ Description—Utilizes a double wall drill pipe and has the reverse circulation of other conventional rotary drilling methods; air or water is forced down the outer casing and is circulated up the inner drill pipe, with the cuttings lifted up to the surface through the inner drill pipe.

    ___ Applications—Used in both unconsolidated and consolidated formations; minimal risk of contamination of sample and/or water-bearing zone from sample collection; in stable formations, wells with diameters as large as 6 inches can be installed in open hole completions; in unstable formations, well diameters are limited to approximately 4 inches.

    ___ Limitations—Air may modify chemical or biological conditions; recovery time is uncertain; filter pack cannot be installed unless the well is a completed open hole.

___ (8) *Driven wells*

    ___ Description—Consist of a steel-well screen that is either welded or attached with drive couplings to a steel casing; well screen and attached casing are forced into the ground by hand, using a weighted drive sleeve or a heavy drive head mounted on a hoist; as well is driven, new sections of casing are attached to the well.

    ___ Applications—Water level monitoring in shallow formations; can be used for water supply wells.

___ Limitations—Depth limited to approximately 50 feet; small diameter casing; steel casing interferes with some chemical analysis; not suitable for dense and/or dry materials.

___ (9) *Jet percussion*

___ Description—Uses a wedge-shaped drill bit attached to the end of the drill pipe; water is forced under pressure down the drill pipe and is discharged through ports on the sides of the drill bit; the drill is lifted and dropped while rotating, and the water is forced up the annular space between the drill pipe and the borehole wall, carrying cutting to the surface.

___ Applications—Primarily used in unconsolidated formations, but may be used in some softer consolidated rock; best application is a 4-inch borehole with a 2-inch casing and screen installed, sealed, and grouted.

___ Limitations—Diameter limited to 4 inches; disturbance of formation possible if borehole is not cased immediately.

___ b. Basic design components of a groundwater-monitoring well (See Fig. 2.F.1.)

___ Well casing—Provides the wall thickness for the stability of the well.

___ Annular space—Area between the well casing and the borehole.

___ Well intake—Provides means for monitoring a particular aquifer.

___ Well screen—Provides mechanism for withdrawing water low in turbidity.

Figure 2.F.1

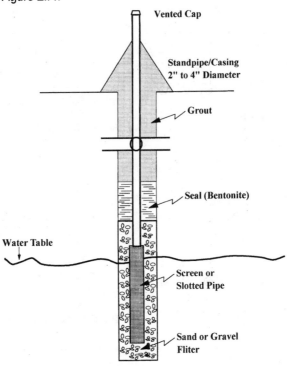

Vented Cap

Standpipe/Casing
2" to 4" Diameter

Grout

Seal (Bentonite)

Water Table

Screen or
Slotted Pipe

Sand or Gravel
Fliter

___ Filter pack—Placed in the annular space between the borehole and the well screen, they minimize the passage of formation materials into the well.

___ Annular seal—Prevents contamination of sample and the groundwater.

___ Surface completion components—Include surface seal, protective casing, ventilation hole, cap and lock.

___ c. Schematic of a detection monitoring system (See Fig. 2.F.2.)

Figure 2.F.2

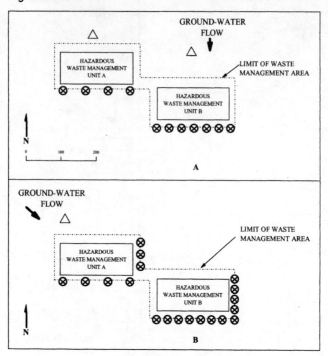

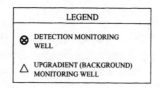

**DOWNGRADIENT WELLS IMMEDIATELY ADJACENT TO THE
HAZARDOUS WASTE MANAGEMENT AREA LIMITS**

___ *d.* Costs of well installation (Ref. 27)
   ___ (1) Test well, 2½-in diameter, 15 to 50
   gpm
      ___ Up to 50 ft deep—$1400
      ___ 75 ft deep—$2000

___ 150 ft deep—$4000
___ 300 ft deep—$6500
___ (2) Borings, std penetration test
    ___ Case borings with samples, 2½-in diameter—$20/boring
    ___ Drilling in rock, with casing and sampling—$40/boring
___ (3) Pumps, 4-in submersible to 100 ft
    ___ ½ hp—$500 to $600
    ___ 1 hp—$600 to $750
    ___ 5 hp—$1500 to $1900
___ (4) Corrosion-resistant pipe, installed
    ___ Iron alloy, 1½-in diameter—$30/LF
    ___ Iron alloy, 3-in diameter—$45/LF
    ___ Iron alloy, 4-in diameter—$55/LF
    ___ Plastic, epoxy, 2-in diameter—$13/LF
    ___ Plastic, epoxy, 4-in diameter—$20/LF
    ___ Plastic, epoxy, 6-in diameter—$27/LF
    ___ Polyester, fiberglass, 2-in diameter—$18/LF
    ___ Polyester, fiberglass, 4-in diameter—$25/LF
    ___ Polyester, fiberglass, 6-in diameter—$36/LF
    ___ Polypropylene, acid resistant, 2-in diameter—$9/LF
    ___ Polypropylene, acid resistant, 3-in diameter—$12/LF
    ___ Polypropylene, acid resistant, 4-in diameter—$15/LF
___ *e.* Field boring log information checklist
    ___ (1) General
        ___ Project name
        ___ Hole name/number
        ___ Date started and finished
        ___ Geologist's name
        ___ Driller's name
        ___ Sheet number

___ Hole location: map and elevation
___ Rig type: bit size/auger size
___ Petrologic-lithologic classification scheme used
___ (2) Information specific to drilling (columns)
___ Depth
___ Sample location/number
___ Blow counts and advance rate
___ Percent sample recovery
___ Narrative description
___ Depth to saturation
___ (3) Geologic observations
___ Soil/rock type
___ Color and stain
___ Gross petrology
___ Friability
___ Moisture content
___ Degree of weathering
___ Presence of carbonate
___ Fractures
___ Solution cavities
___ Bedding
___ Discontinuities
___ Water-bearing zones
___ Formational strike and dip
___ Fossils
___ Depositional structures
___ Organic content
___ Odor
___ Suspected contamination
___ (4) Drilling observations
___ Loss of circulation
___ Advance rates
___ Rig chatter
___ Water levels
___ Amount of air used, air pressure

___ Drilling difficulties

___ Changes in drilling method or equipment

___ Readings from detective equipment, if any

___ Amount of water yield or loss during drilling at different depths

___ Amounts and types of liquids used

___ Running sands

___ Caving/hole stability

___ (5) Additional remarks

___ Equipment failures

___ Possible contamination

___ Deviations from drilling plan

___ Weather

___ *f.* Monitoring well documentation checklist

___ (1) Well design

___ Length, schedule and diameter of casing

___ Joint type (threaded, flush, or solvent-welded)

___ Length, schedule, and diameter of screen

___ Percentage of open area in screen

___ Slot size of screen

___ Distance the filter pack extends above the screen

___ Elevations of the top of well casing, bottom and top of protective casing, ground surface, bottom of borehole, bottom of well screen, and top and bottom of seal(s)

___ Well location by coordinate or grid systems

___ Well location on plan sheet showing the coordinate system, scale, a north arrow, and a key

___ (2) Materials
  ___ Casing and screen
  ___ Filter pack (including grain size analysis)
  ___ Seal and physical form
  ___ Slurry or grout mix used
___ (3) Installation
  ___ Drilling method
  ___ Drilling fluid, as applicable
  ___ Source of water and analysis of water, as applicable
  ___ Time period between the addition of backfill and construction of well protection
___ (4) Development
  ___ Date, time, elevation of water level prior to and after development
  ___ Method used for development
  ___ Time spent developing a well
  ___ Volume of water removed
  ___ Volume of water added, source of water added, chemical analysis of water added, as applicable
  ___ Clarity of water before and after development
  ___ Amount of sediment present at the bottom of the well
  ___ pH, specific conductance, and temperature readings
___ (5) Soils information
  ___ Soil sample test results
  ___ Driller's observation or photocopied driller's log
___ (6) Other information
  ___ Water levels and dates
  ___ Water yield
  ___ Any changes made in well construction, casing elevation, etc.

___ Well identification number

___ Formation samples (depth and method of collection)

___ Water samples (depth, method of collection, and results)

___ Filter pack (depth, thickness, grain size analysis, placement method, supplier)

___ Date(s) of all work

___ Name, address of consultant, drilling company, and stratigraphic log preparer(s)

___ Description and results of pump or stabilization test if performed

___ Methods used to decontaminate drilling equipment and well construction material

___ (7) As-built construction information

___ Top of ground surface

___ Protective grouting and grading at ground surface

___ Well-casing length and depth

___ Screen length and depth

___ Location and extent of gravel pack

___ Location and extent of bentonite seal

___ Water table

___ Earth materials stratigraphy throughout boring

___ Details of bedrock seal for rock wells

___ Depths of water-bearing fractures, faults, or fissures and approximate yield for rock wells

___ 3. Selected parameters from Appendix IX to 40 CFR Part 264, including Chemical, suggested analytical method, and practical quantitation limit—RCRA groundwater monitoring list (See Table 2.F.1.)

TABLE 2.F.1   **Selected Parameters from Appendix IX of 40 CFR 264**

Constituent Name	Suggested Analytical Method	Practical Quantitation Limit (μg/L)
Acetone	EPA Method 8240	100
Acrylonitrile	EPA Method 8230	5
Aldrin	EPA Methods 8240 8080 8270	5 0.05 10
Benzene	EPA Method 8020	2
Cadmium	EPA Methods 6010 7130 7131	40 50 1
1,1-Dichloroethane	EPA Methods 8010 8240	1 5
Fluorene	EPA Methods 8100 8270	200 10
Isobutyl alcohol	EPA Method 8015	50
Methyl ethyl ketone	EPA Methods 8015 8240	10 100
Naphthalene	EPA Methods 8100 8270	200 10
Polychlorinated biphenyls (PCBs)	EPA Methods 8080 8250	50 100
Phenol	EPA Methods 8040 8270	1 10
Sulfide	EPA Method 9030	10,000
1,1,1-Trichloroethane	EPA Method 8240	5
Zinc	EPA Methods 6010 7950	20 50

___ 4. In situ soil analysis methods

    ___ *a. Detectors used in field for soil-gas analysis*

        ___ Flame ionization detector—Can detect total hydrocarbons to a level of 4 picograms/sec of carbon

        ___ Electron capture detector—Can detect halogenated hydrocarbons to a level of 0.01 picograms/sec of material (typically chlorine or fluorine)

        ___ Hall electrolytic conductivity detector— Can detect halogenated species, nitrogen-containing organics, and sulfur-containing organics to a level of 0.5 picrograms/sec (chlorine)

        ___ Flame photometric detector—Can detect sulfur and phosphorus compounds at a level of 1 picogram/sec of phosphorous or 100 picograms/sec of sulfur

        ___ Photoionization detector—Can detect aromatic hydrocarbons to a level of 10 to 100 picograms/sec of aromatic material

    ___ *b. Passive soil–gas sampling badge and manifold* (See Fig. 2.F.3.)

    ___ *c. Soil moisture sampling device* (See Fig. 2.F.4.)

Figure 2.F.3

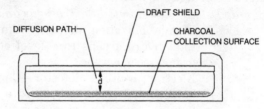

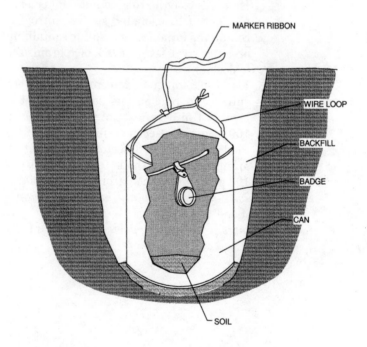

Figure 2.F.4

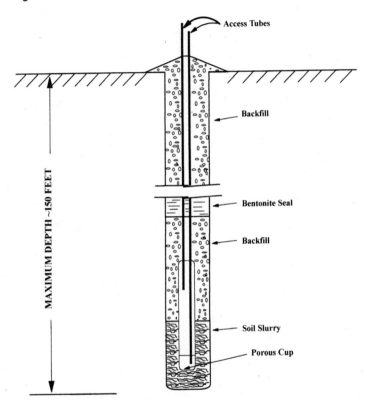

NOTES

___ G. Air Emissions Assessment

___ 1. EPA test methods and applications for sampling toxic organic compounds in ambient air (Ref. 37)

    ___ *a. Method T01*

        ___ Description—Tenax gas chromatograph (GC) adsorption and gas chromatograph/ mass spectrometer (GC/MS) analysis

        ___ Types of compounds determined— Volatile, nonpolar organics having boiling points in the range of 80 to 200°C

        ___ Examples of specific compounds determined—Benzene, carbon tetrachloride, chlorobenzene, chloroform, chloroprene, 1.4-dichlorobenzene, ethylene dichloride, methyl chloroform, nitrobenzene, perchloroethylene, toluene, trichloroethylene, *o,m,p*-xylene

    ___ *b. Method T02*

        ___ Description—Carbon molecular sieve adsorption and GC/MS analysis

        ___ Types of compounds determined—Highly volatile, nonpolar organics having boiling points in the range of –15 to 120°C

        ___ Examples of specific compounds determined—Acrylonitrile, allyl chloride, benzene, carbon tetrachloride, chloroform, ethylene dichloride, methyl chloroform, methylene chloride, toluene, trichloroethylene, vinyl chloride, vinylidene chloride

    ___ *c. Method T03*

        ___ Description—Cryogenic trapping and GC/FID analysis

        ___ Types of compounds determined—Volatile, nonpolar organics having boiling points in the range of –10 to 200°C

        ___ Examples of specific compounds determined—Acrylonitrile, allyl chloride, benzene, carbon tetrachloride, chlorobenzene, chloroform, 1,4-dichlorobenzene, ethylene dichloride, methyl chloroform, methylene chloride, nitrobenzene, perchloroethylene, toluene, trichloroethylene, vinyl chloride, vinylidene chloride, *o,m,p*-xylene

___ d. *Method T04*

    ___ Description—High-volume PUF sampling and GC/ECD analysis

    ___ Types of compounds determined—Organochlorine pesticides and PCBs

    ___ Examples of specific compounds determined—4,4'-DDE, 4,4'-DDT

___ e. *Method T05*

    ___ Description—Dinitrophenylhydrazine liquid impinger sampling and HPLC/UV analysis

    ___ Types of compounds determined—Aldehydes and ketones

    ___ Examples of specific compounds determined—Acetaldehyde, acrolein, benzaldehyde, formaldehyde

___ f. *Method T06*

    ___ Description—HPLC analysis

    ___ Compound determined—Phosgene

___ g. *Method T07*

    ___ Description—Thermosorb/*n* adsorption

    ___ Compound determined—*N*-nitrosodimethylamine

___ h. *Method T08*

    ___ Description—Sodium hydroxide liquid impinger with high performance liquid chromatography

    ___ Compounds determined—Cresol, phenol

___ i. *Method T09*

    ___ Description—High-volume PUF sampling with GC/ECD analysis

    ___ Compound determined—Dioxin

___ *j. Method T10*

    ___ *Description*—Low-volume PUF sampling with GC/EDC analysis

    ___ *Types of compounds determined*—Pesticides

    ___ *Examples of specific compounds determined*—Alochlor 1242, 1254, and 1260, captan, chlorothalonil, chlorpyrifos, dichlorovos, dicofol, dieldrin, endrin, endrin aldehyde, folpet, heptachlor, heptachlor epoxide, hexachlorobenzene, *a*-hexachlorocyclohexane, hexachloropentadiene, lindane, methoxychlor, mexacarbate, mirex, trans-nonachlor, oxychlordane, pentachlorobenzene, 1,2,3-trichlorobenzene, 2,4,5,-trichlorophenol

___ *k. Method T011*

    ___ Description—Adsorbent cartridge followed by HPLC detection

    ___ Types of compounds determined—Aldehydes

    ___ Examples of specific compounds determined—acetaldehyde, acrolein, butyraldehyde, crotomaldehyde, 2,5-dimethylbenzaldehyde, formaldehyde, hexanaldehyde, isovaleraldehyde, propionaldhyde, *o*-toluene, *m*-toluene, *p*-toluene, valeraldehyde

___ *l. Method T012*

    ___ *Description*—Cryogenic PDFID

    ___ Compounds determined—Nonmethane organic compounds

___ *m. Method T013*

    ___ Description—PUF/XAD-2 adsorption with GC and HPLC detection

    ___ Types of compounds determined—Polynuclear aromatic hydrocarbons

    ___ Examples of specific compounds determined—Acenaphthene, acenaphthylene, anthracene, benzo(*a*)anthracene, benzo-(*a*)pyrene, benzo(*b*)fluoranthene, chrysene, dibenzo(*a,h*)anthracene, fluoranthene, fluorene, indeno(1,2,3-*cd*)pyrene, naphthalene, phenanthrene, pyrene

___ *n. Method T014*

    ___ Description—Summar® passivated canister sampling with gas chromatography

    ___ Types of compounds determined—Semivolatile and volatile organic compounds

    ___ Examples of specific compounds determined—Acenaphthene, acenaphthylene, benzene, benzyl chloride, carbon tetrachloride, chlorobenzene, chloroform, 1,2-dichlorobenzene, 1,2-dichloropropane, ethyl chloride, methyl chloroform, perchloroethylene, trichloroethylene, vinyl benzene, vinyl chloride, *o,m,p*-xylene

___ 2. EPA sampling methods for stack sampling (Ref. 6)

___ Method 1—Sample and velocity traverses for stationary sources

___ Method 1A—Sample and velocity traverses for stationary sources with small stacks or ducts

___ Method 2—Determination of stack gas velocity and volumetric flow rate (Type S pitot tube)

___ Method 2A—Direct measurement of gas volume through pipes and small ducts

___ Method 2B—Determination of exhaust gas volume flow rate from gasoline vapor incinerators

___ Method 2C—Determination of stack gas velocity and volumetric flow rate in small stacks or ducts (standard pitot tube)

___ Method 2D—Measurement of gas volumetric flow rates in small pipes and ducts

___ Method 2E—Determination of landfill gas; gas production flow rate

___ Method 3—Gas analysis for carbon dioxide, oxygen, excess air, and dry molecular weight

___ Method 3A—Determination of oxygen and carbon dioxide concentration in emissions from stationary sources (instrumental analyzer procedure)

___ Method 4—Determination of moisture content in stack gases

___ Method 5—Determination of particulate emissions from stationary sources

___ Method 5A—Determination of particulate emissions from the asphalt processing and asphalt roofing industry

___ Method 5B—Determination of nonsulfuric acid particulate matter from stationary sources

___ Method 5C—[Reserved]

___ Method 5D—Determination of particulate emissions from positive pressure fabric filters

___ Method 5E—Determination of particulate emissions from the wool fiberglass insulation–manufacturing industry

___ Method 5F—Determination of nonsulfate particulate matter from stationary sources

___ Method 5G—Determination of particulate emissions from wood heaters from a dilution tunnel sampling location

___ Method 5H—Determination of particulate emissions from wood heaters from a stack location

___ Method 6—Determination of sulfur dioxide emissions from stationary sources

___ Method 6A—Determination of sulfur dioxide, moisture, and carbon dioxide emissions from local fossil fuel combustion sources

___ Method 6B—Determination of sulfur dioxide and carbon dioxide daily average emissions from fossil fuel combustion sources

___ Method 6C—Determination of sulfur dioxide emissions from stationary sources (instrumental analyzer procedure)

___ Method 7—Determination of nitrogen oxide emissions from stationary sources

___ Method 7A—Determination of nitrogen oxide emissions from stationary sources—ion chromatographic method

___ Method 7B—Determination of nitrogen oxide emissions from stationary sources (ultraviolet spectrophotometry)

___ Method 7C—Determination of nitrogen oxide emissions from stationary sources—alkaline-permanganate/colorimetric method

___ Method 7D—Determination of nitrogen oxide emissions from stationary sources—alkaline-permanganate/ion chromatographic method

___ Method 7E—Determination of nitrogen oxides emissions from stationary sources (instrumental analyzer procedure)

___ Method 8—Determination of sulfuric acid mist and sulfur dioxide emissions from stationary sources

___ Method 9—Visual determination of the opacity of emissions from stationary sources (Alternate method—determination of the opacity of emissions from stationary sources remotely by lidar)

___ Method 10—Determination of carbon monoxide emissions from stationary sources

___ Method 10A—Determination of carbon monoxide emissions in certifying continuous emission-monitoring systems at petroleum refineries

___ Method 10B—Determination of carbon monoxide emissions from stationary sources

___ Method 11—Determination of hydrogen sulfide content of fuel gas streams in petroleum refineries

___ Method 12—Determination of inorganic lead emissions from stationary sources

___ Method 13A—Determination of total fluoride emissions from stationary sources—SPADNS zirconium lake method

___ Method 13B—Determination of total fluoride emissions from stationary sources—specific ion-electrode method

___ Method 14—Determination of fluoride emissions from potroom roof monitors for primary aluminum plants

___ Method 15—Determination of hydrogen sulfide, carbonyl sulfide, and carbon disulfide emissions from stationary sources

___ Method 15A—Determination of total reduced sulfur emissions from sulfur recovery plants in petroleum refineries

___ Method 16—Semicontinuous determination of sulfur emissions from stationary sources

___ Method 16A—Determination of total reduced sulfur emissions from stationary sources (impinger technique)

___ Method 16B—Determination of total reduced sulfur emissions from stationary sources

___ Method 17—Determination of particulate emissions from stationary sources (in-stack filtration method)

___ Method 18—Measurement of gaseous organic compound emissions by gas chromatography

___ Method 19—Determination of sulfur dioxide removal efficiency and particulate, sulfur dioxide, and nitrogen oxides emission rates

___ Method 20—Determination of nitrogen oxides, sulfur dioxide, and diluent emissions from stationary gas turbines

___ Method 21—Determination of volatile organic compound leaks

___ Method 22—Visual determination of fugitive emissions from material sources and smoke emissions from flares

___ Method 23—Determination of polychlorinated dibenzo-*p*-dioxins and polychlorinated dibenzo-furans from stationary sources

___ Method 24—Determination of volatile matter content, water content, density, volume solids, and weight solids of surface coatings

___ Method 24A—Determination of volatile matter content and density of printing inks and related coatings

___ Method 25—Determination of total gaseous non-methane organic emissions as carbon

___ Method 25A—Determination of total gaseous organic concentration, using a flame ionization analyzer

___ Method 25B—Determination of total gaseous organic concentration, using a nondispersive infrared analyzer

___ Method 26—Determination of hydrogen chloride emissions from stationary sources

___ Method 27—Determination of vapor tightness of gasoline delivery tank, using pressure-vacuum test

___ Method 28—Certification and auditing of wood heaters

___ Method 28A—Measurement of air-to-fuel ratio and minimum achievable burn rates for wood-fired appliances

___ 3. Example stack sampling trains

    ___ *Organic Concentration Measurement System* (See Fig. 2.G.1.)

Figure 2.G.1

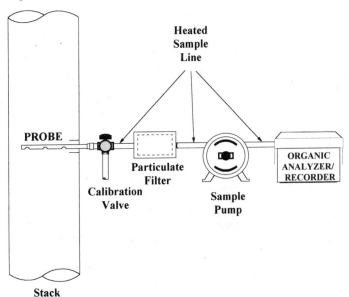

Grab Sampling Train (See Fig. 2.G.2.)
Nitrogen Dioxide Sampling Train (See Fig. 2.G.3.)
Moisture Sampling Train—Reference Method (See Fig. 2.G.4.)

Figure 2.G.2

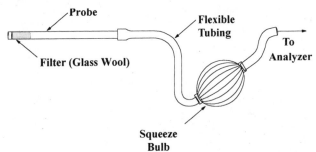

Figure 2.G.3

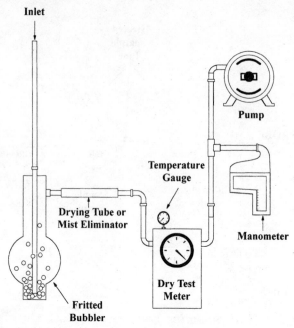

Inlet

Drying Tube or
Mist Eliminator

Fritted
Bubbler

Temperature
Gauge

Dry Test
Meter

Pump

Manometer

Figure 2.G.4

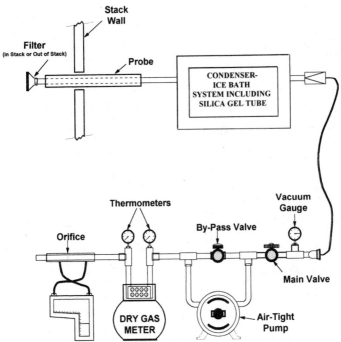

NOTES

PART 3

Occupational Health Management

NOTES

___ A. Chemical Hazards

___ 1. Major classes of common organic solvents (Refs. 24 and 33)

    ___ a. *Aliphatic hydrocarbons (acyclic)*

        ___ (1) Chemical structure—Has straight or branched chains of the constituent carbon; graphic of the chemical structure of *n*-hexane (See Fig. 3.A.1.)

Figure 3.A.1

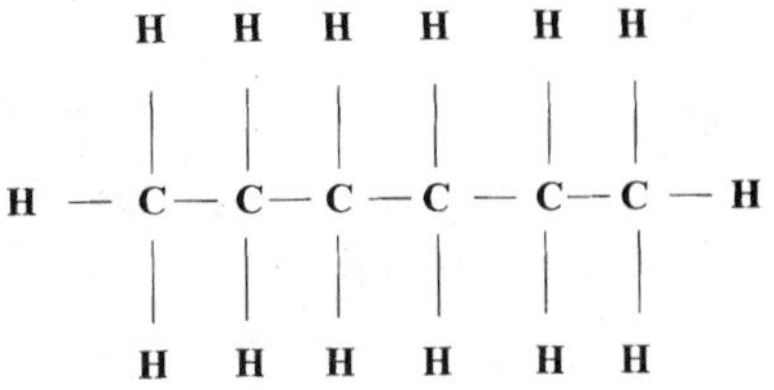

Hexane

        ___ (2) Three subgroups

          ___ Paraffins, or alkanes—Saturated and comparatively unreactive

          ___ Olefines, or alkenes or alkadienes—Unsaturated and quite reactive

          ___ Acetylenes, or alkynes—Contain a triple bond; highly reactive; may cause dermatitis

        ___ (3) Examples—Ethylene, isobutane, *n*-hexane, hexane, octane, 2-methylpentane and 2,2-dimethylpentane, 1-octene, 1-octadecene, methane

        ___ (4) Main Hazards—Causes dermatitis; can act as a depressant to central nervous system

    ___ b. *Aromatic hydrocarbons*

        ___ (1) Chemical structure—An unsaturated cyclic hydrocarbon containing one or more rings; graphic of the chemical structure of benzene (See Fig. 3.A.2.)

Figure 3.A.2

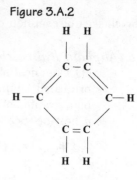

Benzene

___ (2) Characteristics and other information

    ___ The compounds of this group are derived mainly from petroleum and coal tar.

    ___ They tend to be highly reactive and chemically versatile.

    ___ The name indicates a strong (though not unpleasant) odor characteristic of most substances of this nature.

___ (3) Examples—Toluene, xylene, benzene

___ (4) Main hazards—Benzene negatively affects blood-forming tissues of the bone marrow; other aromatic hydrocarbons cause dermatitis and act as a depressant to the central nervous system.

___ c. *Cyclic hydrocarbons (cyclopariffins)*

    ___ (1) Chemical structure—Has a ring structure saturated and unsaturated with hydrogen; graphic of the chemical structure of cyclohexane (a cyclic hydrocarbon) (See Fig. 3.A.3.)

    ___ (2) Characteristics and other information

        ___ May be mono-, di-, tri-, or poly-ringed.

Figure 3.A.3

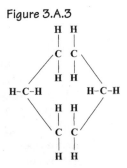

Cyclohexane

____ The unsaturated cyclic hydrocarbons
are generally more irritating than
the saturated forms.

____ (3) Examples—Cyclopropane, cyclobutane,
cyclopentane, cyclohexane, turpentine

____ (4) Main hazards—Slightly toxic; irritants

____ d. *Halogenated hydrocarbons*

____ (1) Chemical structure—A halogen atom
(fluorine, chlorine, bromine, or iodine)
has replaced one or more hydrogen atoms
on the hydrocarbon; a graphic of the
chemical structure of carbon tetrachloride
(See Fig. 3.A.4.)

Figure 3.A.4

Carbon tetrachloride

____ (2) Characteristics and other information

____ Generally inert, nonflammable, with
a wide range of solvency.

___ Some are considered ozone depleting.

___ Some may react violently with reactive substances such as barium, sodium, and potassium.

___ (3) Examples—Chlorofluorocarbons such as CFC 113 and CFC 121, carbon tetrachloride, 1,1,1-trichloroethane, trichlorotrifluoroethane, chloroform, chlorobenzene, trifluoromethane

___ (4) Main hazards—Many are highly toxic; have synergistic effects with alcohol ingestion

___ e. Esters

___ (1) Chemical structure—Typically formed by interaction of an organic acid with an alcohol or other organic compounds rich in OH groups; graphic of the chemical structure ethyl acetate (See Fig. 3.A.5.)

Figure 3.A.5

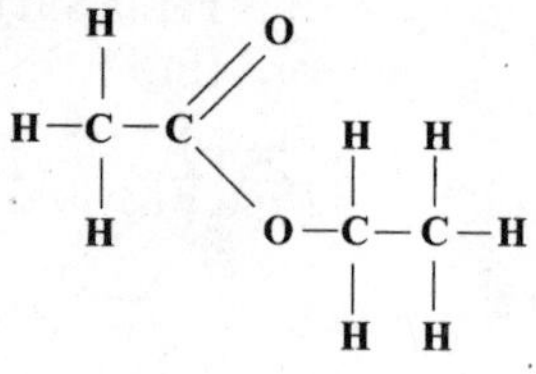

Ethyl acetate

___ (2) Characteristics and other information

___ Esters of acetic acid are called *acetates*.

___ Esters of carbonic acid are called *carbonates*.

___ Properties are partly determined by the parent alcohol.

___ Generally, esters are good solvents for surface coatings.

___ (3) Examples—Amyl acetate, ethyl acetate, diethyl carbonate, diphenyl carbonate

___ (4) Major hazards—Tend to have irritating effects on exposed skin surfaces and on the respiratory tract; they are potent anesthetics.

___ f. *Ethers*

___ (1) Chemical structure—Have an oxygen atom interposed between two carbon atoms (organic groups) in the molecular structure; graphic of the chemical structure of ethyl ether (See Fig. 3.A.6.)

Figure 3.A.6

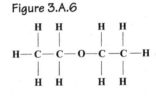

Ethyl ether

___ (2) Characteristics and other information

___ Ethers are typically derived from catalytic hydration of olefins, although they may also be derived from alcohols by elimination of water.

___ They are often found in liquid form.

___ They are generally chemically inert, stable, and easy to recover in extraction processes.

___ Lower molecular-weight ethers present fire and explosion hazards.

___ (3) Examples—Isopropyl ether, ethylene glycol monomethyl ether, ethyl ether

___ (4) Main hazard—Have a tendency to form explosive peroxides

___ g. *Alcohols*

    ___ (1) Chemical structure—Types and nomenclature

      ___ Monohydric—One OH group and termed *alcohols*

      ___ Dihydric—Two OH groups and termed *glycols*

      ___ Trihydric—Three OH groups and termed *glycerols*

      ___ Polyhydric—Three or more OH groups and termed *polyols*

      ___ Graphic of the chemical structure of a monohydric alcohol, methanol (See Fig. 3.A.7.)

Figure 3.A.7

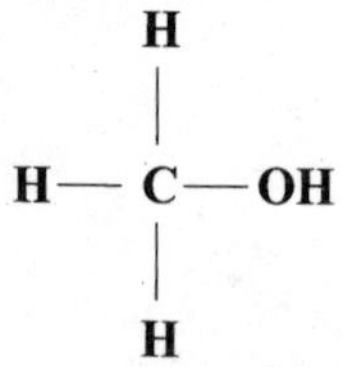

$$H-\overset{\displaystyle \overset{H}{|}}{\underset{\displaystyle \underset{H}{|}}{C}}-OH$$

Methanol

      ___ Graphic of a dihydric alcohol, ethylene glycol (See Fig. 3.A.8.)

Figure 3.A.8

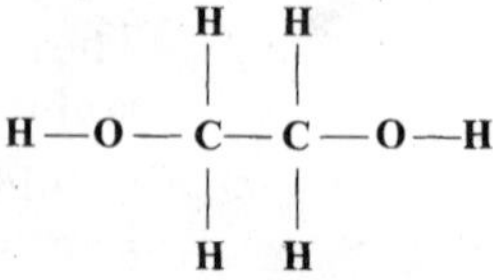

$$H-O-\overset{\displaystyle \overset{H}{|}}{\underset{\displaystyle \underset{H}{|}}{C}}-\overset{\displaystyle \overset{H}{|}}{\underset{\displaystyle \underset{H}{|}}{C}}-O-H$$

Ethylene glycol

___ (2) Characteristics and other information
  ___ Hydroxyl-containing organic compounds occur naturally in plants and are made synthetically from petroleum derivatives.
  ___ Typically used as solvents.
  ___ Vary widely in toxicity.
___ (3) Examples
  ___ Examples of monohydrics include alkyl alcohol, ethanol, and methanol.
  ___ An example of a dihydric is ethylene glycol.
  ___ An example of a trihydric is glycerin.
  ___ Examples of polyols with more than three OH groups are sugar alcohols.
___ (4) Main hazards—Can affect central nervous system and the liver; some are highly toxic.
___ *h. Aldehydes*
  ___ Chemical structure—Aldehydes are characterized by an unsaturated carbonyl group, C=O. Graphic of the chemical structure of acetaldehyde (See Fig. 3.A.9.)

Figure 3.A.9

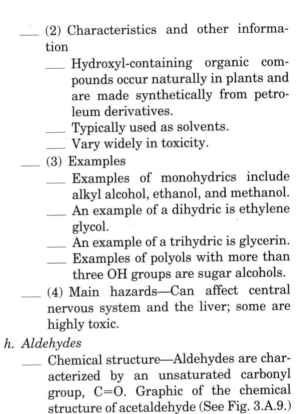

Acetaldehyde

  ___ Other information—Formed from alcohols by either dehydrogenation or oxidation, these compounds occupy an intermediate position between primary alcohols and the acids obtained from them by further oxidation.

___ Examples—Aldehyde ammonia, acetaldehyde

___ Main Hazards—May cause skin and mucosal irritation; may affect central nervous system; may have sensitizing properties

___ *i. Ketones*

___ (1) Chemical structure—Ketones are characterized by a carboxyl group, C=O, attached to two alkyl groups; graphic of the chemical structure of methyl ethyl ketone (See Fig. 3.A.10.)

Figure 3.A.10

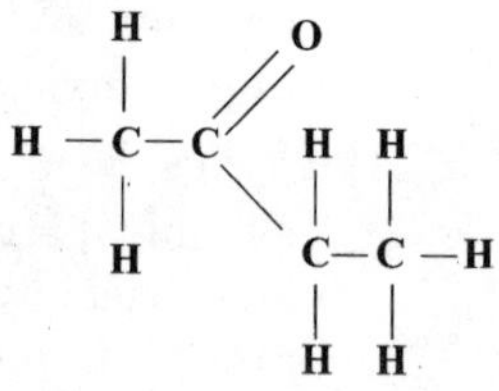

Methyl ethyl ketone

___ (2) Characteristics and other information

___ Derived by oxidation of secondary alcohols

___ Excellent as solvents; generally stable, with high-dilution rations for hydrocarbon diluents

___ (3) Examples—Acetone, diethyl ketone, methyl ethyl ketone

___ (4) Main hazard—May be irritating to eyes, nose, and throat

___ 2. Classes of other chemicals—Health hazards
___ *a.* *Carcinogens*—Chemicals that have the potential to cause cancer (Refs. 7 and 9)
___ (1) Designations—Abbreviations for agency codings
___ EPA—Environmental Protection Agency (U.S.)
___ IARC—International Agency for Research on Cancer
___ MAK—Federal Republic of Germany Maximum Concentration Values in the Workplace
___ NIOSH—National Institute for Occupational Safety and Health
___ NTP—National Toxicology Program
___ OSHA—Occupational Safety and Health Association
___ (2) Designations by agency
___ EPA A—Human carcinogen: sufficient evidence from epidemiologic studies to support a causal association between exposure and cancer
___ EPA B—Probable human carcinogen: weight of evidence of carcinogenicity based on epidemiologic studies is limited; agents for which weight of evidence of carcinogenicity is animal study–based are sufficient. Two subgroups:
___ B1—Limited evidence of carcinogenicity from epidemiologic studies
___ B2—Sufficient evidence from animal studies; inadequate evidence or no data from epidemiologic studies
___ EPA C—Possible human carcinogen: limited evidence of carcinogenicity in

animals in the absence of human data

___ EPA D—Not classifiable as to human carcinogenicity: inadequate human and animal evidence of carcinogenicity, or no data are available

___ EPA E—Evidence of noncarcinogenicity for humans: no evidence for carcinogenicity in at least two adequate animal tests in different species or in both adequate epidemiologic and animal studies

___ IARC-1—Carcinogenic to humans: sufficient evidence of carcinogenicity

___ IARC-2A—Probably carcinogenic to humans: limited human evidence; sufficient evidence in experimental animals

___ IARC-2B—Possibly carcinogenic to humans: limited evidence in humans in the absence of sufficient evidence in experimental animals

___ IARC-3—Not classifiable as to carcinogenicity to humans

___ IARC-4—Probably not carcinogenic to humans

___ MAK-A1—Capable of inducing malignant tumors as shown by experience with humans

___ MAK-A2—Unmistakably carcinogenic in animal experimentation only

___ MAK-B—Justifiably suspected of having carcinogenic potential

___ NIOSH X—Carcinogen defined with no further categorization

___ NTP-1—Known to be a carcinogen; sufficient evidence of carcinogenicity from studies in humans, which indicates a causal relationship between the agent and human cancer

___ NTP-2—Reasonably anticipated to be a carcinogen

    ___ A. Limited evidence of carcinogenicity from studies in humans, which indicates that causal relationship is credible

    ___ B. Sufficient evidence of carcinogenicity from studies in experimental animals

___ OSHA-X—Carcinogen defined with no further categorization

___ TLV-A1—Confirmed human carcinogen: Agent is carcinogenic to humans based on epidemiologic studies of, or convincing clinical evidence in, exposed humans.

___ TLV-A2—Suspected human carcinogen: Agent is carcinogenic in experimental animals at dose levels, by route(s) of administration, at site(s), of histologic type(s), or by mechanism(s) considered relevant to worker exposure; available epidemiologic studies are conflicting or insufficient to confirm an increased risk of cancer in exposed humans.

___ TLV-A3—Animal carcinogen: Agent is carcinogenic in experimental animals at a relatively high dose, by route(s) of administration, at site(s), in histologic type(s), or by mechanism(s) not considered relevant to worker exposure; available epidemiologic studies do not confirm an increased risk of cancer in exposed humans; available evidence suggests that the agent is not likely to cause cancer in humans except under uncommon or unlikely routes or levels of exposure.

___ TLV-A4—Not classifiable as a human carcinogen: Inadequate data on which to classify the agent in terms of its carcinogenicity in humans and/or animals

___ TLV-A5—Not suspected as a human carcinogen: Not suspected to be a human carcinogen on the basis of properly conducted epidemiologic studies in humans; studies have sufficiently long follow-up, reliable exposure histories, sufficiently high dose, and adequate statistical power to conclude that exposure to the agent does not convey a significant risk of cancer to humans; evidence suggesting a lack of carcinogenicity in experimental animals will be considered if it is supported by other relevant data; substances for which no human or experimental animal–carcinogenic data have been reported are assigned no carcinogen designation; exposures to carcinogens must be kept to a minimum; workers exposed to A1 carcinogens without a TLV should be properly equipped to eliminate to the fullest extent possible all exposure to the carcinogen; for A1 carcinogens with TLV and for A2 and A3 carcinogens, worker exposure by all routes should be carefully controlled to levels as low as reasonably achievable below the TLV.

___ (3) Examples of known or suspected carcinogens—Asbestos, benzene, lead, carbon tetrachloride, vinyl chloride

___ *b. Highly toxic materials*

   ___ (1) Definition

     ___ The chemical has a median lethal dose (LD_{50}) of 50 milligrams or less per kilogram (50 mg/kg) of body weight when administered orally to albino rats weighing between 200 and 300 grams (g) each.

     ___ The chemical has a median lethal dose (LD_{50}) of 200 milligrams or less per kilogram (200 mg/kg) of body weight when administered by continuous contact for 24 hours (or less if death occurs before 24 hours) with the bare skin of albino rabbits weighing between 2 and 3 kilograms (kg) each.

     ___ The chemical has a median lethal concentration (LC_{50}) in air of 200 parts per million (ppm) by volume or less of gas or vapor, or 2 milligrams per liter (2 mg/L) or less of mist, fume or dust, when administered by continuous inhalation for 1 hour (or less if death occurs before 1 hour) to albino rats weighing between 200 and 300 grams each.

   ___ (2) Examples—Arsine, nitrogen dioxide, nicotine, acrolein, pentachlorophenol

___ *c. Toxic materials*

   ___ (1) Definition

     ___ The chemical has a median lethal dose (LD_{50}) of more than 50 milligrams per kilogram (50 mg/kg) but not more than 500 milligrams per kilogram (500 mg/kg) of body weight when administered orally to albino rats weighing between 200 and 300 grams each.

___ The chemical has a median lethal dose (LD_{50}) of more than 200 milligrams per kilogram (200 mg/kg) but not more than 1000 milligrams per kilogram (1000 mg/kg) of body weight when administered by continuous contact for 24 hours (or less if death occurs before 24 hours) with the bare skin of albino rabbits weighing between 2 and 3 kilograms each.

___ The chemical has a median lethal concentration (LC_{50}) in air of more than 200 parts per million but not more than 2000 parts per million by volume of gas or vapor, or 2 milligrams per liter (2 mg/L) or less of mist, fume or dust, when administered by continuous inhalation for 1 hour (or less if death occurs before 1 hour) to albino rats weighing between 200 and 300 grams each.

___ (2) Examples—Chlorine, phosgene, acrylonitrile, allyl alcohol, o-cresol, barium chloride, phenol

___ d. Corrosive materials

___ (1) Definition—Causes a visible destruction of living tissue

___ (2) Examples

___ Acids—Chromic acid, formic acid, sulfuric acid (>4%)

___ Bases—Potassium carbonate, calcium hydroxide, ammonium hydroxide (>10%)

___ Other corrosives—bromine, fluorine, chlorine, ammonia

___ e. *Irritants*

___ (1) Definition—A chemical that can cause inflammation of the mucous membrane of the respiratory tract

___ (2) Examples—Chloracetophene, xylyl bromide

___ f. *Sensitizers*

___ (1) Definition—Causes a large number of exposed people or animals to develop an allergic reaction in normal tissue after repeated exposure to the chemical; effects typically are reversible once the exposure ceases.

___ (2) Examples—Fiberglass dust, epichlorohydrin

___ g. *Asphyxiants*

___ (1) Definition—Inert elements that, in sufficient quantity, exclude oxygen from the body

___ (2) Examples—Carbon monoxide, cyanides

___ h. *Anesthetics*

___ (1) Definition—Chemicals which act to depress the central nervous system

___ (2) Examples—Alcohol, acetylene hydrocarbons, ethyl ether

___ i. *Hepatotoxic agents*

___ (1) Definition—Chemicals that damage the normal functioning of the liver

___ (2) Examples—Carbon tetrachloride, nitrosamines, tetrachloroethane

___ j. *Nephrotoxic agents*

___ (1) Definition—Chemicals that damage the nervous system

         __ (2) Examples—methyl mercury, tetra-ethyl lead

__ *k. Blood damaging agents*

         __ (1) Definition—Chemicals that break down the red blood cells or chemically affect the hemoglobin in the blood

         __ (2) Examples—Benzene, arsine, aniline

__ 3. Other classes of chemicals—Physical hazards

    __ *a. Reactive materials*

         __ (1) Definition—Unstable materials, or materials that react violently when exposed to water or air

         __ (2) Examples—Picric acid, TNT, calcium carbide

    __ *b. Combustible liquid*

         __ (1) Definition—Has a flash point at or above 100°F but below 200°F

         __ (2) Examples—Fuel oil, crude oil

    __ *c. Flammable material*

         __ (1) Definition

            __ An aerosol that, when tested by the method described in 16 CFR 1500.45, yields a flame projection exceeding 18 inches at full valve opening or a flashback at any degree of valve opening

            __ A gas that, at ambient temperature and pressure, forms a flammable mixture with air at a concentration of 13% by volume or less

            __ A gas that, at ambient temperature and pressure, forms a range of flammable mixtures with air wider than 12% by volume, regardless of the lower limit

            __ A liquid having a flashpoint below 100°F

____ A solid, other than a blasting agent or explosive, that is likely to cause fire through friction, absorption of moisture, spontaneous chemical change, or retained heat from manufacturing or processing, or that can be ignited readily and when ignited burns so vigorously and persistently as to create a serious hazard

____ A solid that, when tested by the method described in 16 CFR 1500.44, ignites and burns with a self-sustained flame at a rate greater than one-tenth of an inch per second ($\frac{1}{10}$ in/s) along its major axis

____ (2) Examples—Isopropyl alcohol, acetone, gasoline

____ d. *Explosive materials*

    ____ (1) Definition—Materials that cause a sudden, almost instantaneous, release of pressure, gas, and heat when subjected to sudden shock, pressure, or high temperature

    ____ (2) Examples—Nitroglycerine, ammunition, dynamite

____ e. *Pyrophorics*

    ____ (1) Definition—Materials that will ignite spontaneously in air at a temperature of 130°F or below

    ____ (2) Examples—White phosphorus, superheated toluene, lithium hydride, silane gas

____ f. *Oxidizers*

    ____ (1) Definition—Chemicals that initiate or promote combustion in other materials, thereby causing fire either of itself or through the release of oxygen or other gases

_____ (2) Examples—Sodium nitrate, bromine, chromic acid, calcium hypochlorite

_____ g. *Organic peroxides*

    _____ (1) Definition—Oxidizers that are extremely unstable and which can polymerize, decompose, condense, or become self-reactive when exposed to shock, heat, or friction

    _____ (2) Examples—Methyl ethyl ketone peroxide, dibutyl peroxide

_____ 4. Chemical hazard communication

    _____ a. Material safety data sheet (MSDS)

        _____ (1) OSHA-required Information

           _____ Chemical and common names

           _____ Physical and chemical characteristics

           _____ Physical hazards

           _____ Health hazards

           _____ Primary routes of entry

           _____ Permissible exposure limits

           _____ Carcinogen or potential carcinogen

           _____ Safe handling procedures

           _____ Control measures

           _____ Emergency and first aid procedures

           _____ Manufacturer's name and related information

        _____ (2) Additional information

           _____ Emergency overview

           _____ Fire-fighting measures

           _____ Accidental release measures

           _____ Toxicological information

           _____ Ecological information

           _____ Regulatory information

    _____ b. Chemical labels

        _____ Should identify hazardous chemical(s)

        _____ Should contain appropriate hazard warnings

___ Should include name and address of chemical manufacturer, importer, or other responsible party

___ 5. Routes of entry into the body

___ Oral (ingestion)

___ Respiratory

___ Skin

___ Eyes

___ 6. Exposure limits (Ref. 9)

___ a. *Permissible exposure limit (PEL) and time-weighted average (TWA)*

___ Defines the maximum time-weighted exposure over an 8-hour work shift of a 40-hour workweek that should not be exceeded.

___ OSHA-enforceable limits.

___ b. *Threshold limit value–time-weighted average (TLV-TWA)*

___ Defines a concentration to which nearly all workers can be repeatedly exposed over a normal 8-hour workday in a 40-hour workweek without adverse effects.

___ Defined by the American Conference of Governmental Industrial Hygienists (ACGIH) as nonenforceable industry standards.

___ c. *Short-term exposure limit*

___ Defines a 15-minute time-weighted average–exposure limit that should not be exceeded.

___ ACGIH recommends that exposure at this limit should not be repeated more than four times daily and that at least a 60-minute rest period between exposures should be allowed.

___ d. *Ceiling (C) limit*

    ___ Defines a maximum concentration that should not be exceeded for any period of time.

    ___ Established for certain chemicals that, at certain concentrations, may produce acute poisoning during very short exposures.

___ e. *Action limit*

    ___ Defines a concentration at which steps should be taken to control exposure.

    ___ Established by OSHA for particularly hazardous chemicals/materials, such as asbestos and cadmium.

___ f. *Exposure limit calculations*

    ___ $\text{TWA} = \Sigma\, C_i t_i / \Sigma\, t_i$

        where $\Sigma\, C_i t_i =$ sum of all exposure concentrations times ($\times$) the exposure time; and
            $\Sigma\, t_i =$ the sum of all exposure times

    ___ Example of safe exposure (See Fig. 3.A.11.)

Figure 3.A.11

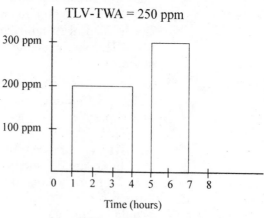

8-hour TWA = 175 ppm

___ Acceptable additive effects $= C_1/T_1 + C_2/T_2 + \cdots C_n/T_n \leq 1$
where $C_1/T_1 + C_2/T_2 + \cdots C_n/T_n =$ the sum of individual chemical exposure concentration divided by ($\div$) the corresponding threshold limit value

___ Examples of calculated additive effects (See Table 3.A.1.)

TABLE 3.A.1 Examples of Additive Effects Calculations

Component A	Mixture ppm	TLV ppm	Component B	Mixture ppm	TLV	Sum	TLV Exceeded?
Methyl ethyl ketone	100	200	Isopropyl alcohol	250	400	1.13	Yes
Methyl ethyl ketone	100	200	Isopropyl alcohol	150	400	0.88	No
Methyl ethyl ketone	75	200	Isopropyl alcohol	275	400	1.1	Yes
Methyl ethyl ketone	75	200	Isopropyl alcohol	175	400	0.81	No

___ 7. Medical surveillance
___ a. Workers required to be in the surveillance program
___ Workers who are routinely exposed to certain concentrations of regulated chemicals or dusts such as lead, benzene, formaldehyde, acrylonitrile, and asbestos
___ Workers who routinely wear a respirator
___ Workers who take part in hazardous waste operations and emergency response activities
___ b. Health questionnaire
___ Occupational history
___ Past medical history
___ Family history
___ Illnesses
___ Smoking habits

___ *c.* Types of tests

___ Physical exam

___ General health survey

___ Blood count

___ Chemistry screening

___ Urinalysis

___ Spirometry test

___ Chest X-ray

___ Audiogram

___ Electrocardiogram

___ *d.* Examples of adopted biological exposure determinants (Ref. 9) (See Table 3.A.2.)

TABLE 3.A.2 **Examples of Adopted Biological Exposure Determinants**

Chemical	Biological Specimen	Sampling Time	Biological Exposure Index*
Acetone	Urine	End of shift	100 mg/L
Cadmium	Urine	Not critical	5 µg/g creatinine
Cadmium	Blood	Not critical	5 µg/L
Methanol	Urine	End of shift	15 mg/L
Parathion	Total *p*-nitrophenol in urine	End of shift	0.5 mg/g creatinine
Parathion	Cholinesterase activity in red cells	Discretionary	70% of individual's baseline

* Biological exposure index (BEI) is the reference value intended as a guideline for the evaluation of potential health hazards in the practice of industrial hygiene. A BEI represents the level of determinants which are most likely to be observed in specimens collected from a healthy worker who has been exposed to chemicals to the same extent as a worker with inhalation exposure to the TLV. The above information is for illustrative purposes. Contact American Conference of Governmental Industrial Hygienists to obtain complete data on BEIs.

___ 8. Odor threshold (Ref. 10)

___ *a.* General

___ Odor is a warning property only and is not to be used as a means of determining chemical exposure.

___ Sensory perception of odorants has four major dimensions: detectability, intensity, character, and hedonic tone (i.e., relative pleasantness or unpleasantness).

___ Detection threshold is the awareness of the presence of an added substance, but not necessarily the recognition of it as an odor sensation.

___ Recognition threshold is the minimum concentration that is recognized typically by 50% of the population as having a characteristic odor quality.

___ b. Odor thresholds in comparison with established health standards (Refs. 9 and 10) (See Table 3.A.3)

TABLE 3.A.3   **Odor Thresholds in Comparison with Established Health Standards**

Chemical name	TWA (ppm)	STEL (ppm)	Odor Threshold (range of accepted values) (ppm)
Acetic acid	10	15	0.037–0.15
Allyl alcohol	2	4	1.4–2.1
Carbon tetrachloride	5	10	140–548
Chloroform	10	—	133–276
Ethyl acetate	400	—	6.4–50
Ethylene dichloride	10	—	6–111
Isopropyl alcohol	400	500	37–610
Methyl ethyl ketone	200	300	2–85
Propylene oxide	20	—	10–199
Toluene	50	—	0.16–37

___ 9. Monitoring for chemical exposure

___ a. *Workplace monitoring—analyzers*

___ (1) Chemiluminescence analyzer—Detects oxides of nitrogen in ppm range

___ (2) Oxygen meter—Reads percentage of oxygen in atmosphere

___ (3) Combustible gas indicator—Used to determine if flammable/combustible material is present in a concentration that could be dangerous

___ (4) Length of stain detector tube—Provides a colorimetric visual indicator for detection of specific chemical

___ (5) Infrared analyzer—Uses a spectrometer to read chemical "fingerprints," or concentration intensities

___ (6) Flame ionization detector—Detects total hydrocarbons in the ppm range

___ (7) Photoionization detector—Measures total concentrations of many organic and some inorganic gases

___ *b.* Personal monitoring

___ (1) Sampling with a sampling pump

___ Standard across industry

___ Charcoal tube adsorption sampler or silica gel tube sampler common adsorption devices

___ (2) Passive (diffusional) sampling

___ No sampling pump used

___ Attached to worker near breathing zone

___ Vapors collected by diffusion on adsorbent such as charcoal

NOTES

NOTES

___ B. Hearing Conservation

___ 1. Sound
 ___ a. Definition
   ___ Sound is produced by any pressure varia-
   tion caused by vibrations.
   ___ Sound travels through any phase—gas,
   liquid, or solid.
   ___ Velocity of sound depends upon the elas-
   ticity and density of the medium.
   ___ Sound pressure is measured in units of
   pressure, such as micropascals (μPa),
   newtons per square meter (N/m^2), micro-
   bars (μbar), and dynes per square centi-
   meter (d/cm^2).
 ___ b. Sound wave
   ___ Sound wave is a longitudinal vibration
   of a conducting medium such as air or
   water.
   ___ Sound waves can be represented as sinu-
   soidal patterns with amplitudes and frequencies.
   ___ Amplitude of sound wave represents
   intensity of sound.
   ___ Loudness is related to intensity.
   ___ Frequency of sound wave is the number
   of times the vibrating object completes its
   cycle of motion in a period of one second.
   ___ Frequency is measured in Hertz (Hz).
   ___ Pitch is related to frequency—the higher
   the pitch, the higher the frequency.
 ___ c. Sound scales
   ___ A-scale—Largely filters out very low frequencies
   ___ B-scale—Moderately filters out very low
   frequencies
   ___ C-scale—Only sightly filters out very low
   frequencies

_____ Noise limits in the OSHA regulation stated in decibels on the A-scale

_____ d. Sound level meters

_____ Measure the sound intensity at a particular moment at a particular place

_____ Can be used at various locations throughout the workplace and at different times of the day in order to estimate overall exposure levels

_____ Specifications for operation and use found in ANSI S1.4-1983

_____ e. Dosimeters

_____ Measure the overall sound exposure directly

_____ Are worn by the employee and measure sound levels wherever the employee travels

_____ Automatically integrate measurements over time

_____ Specifications for operation and use found in ANSI S1.24-1991

_____ f. _Sound pressure calculations_

_____ Scale is logarithmic.

_____ Unit is decibel (dB).

_____ $L_p = 20 \log P/P_0$

where L_p is the sound pressure level of an audible sound.

P_0 is the hearing threshold of an average person at a reference tone of 1000 Hz and sound pressure of 20 µPA.

_____ Decibels increase as sound pressure increases (See Table 3.B.1).

_____ g. _Sound power (energy) calculations_

_____ Scale is logarithmic.

_____ Unit is decibel (dB).

_____ Sound pressure is proportional to the square root of the sound power.

TABLE 3.B.1 **Calculated Increases in Decibels with Increasing Sound Pressure and Sound Energy (Power)**

Increase in Sound Energy (power)	Increase in Decibels	Increase in Sound Pressure	Increase in Decibels
2X	3	2X	6
3X	4.8	3X	9.5
4X	6.0	4X	12.0
5X	7.0	5X	14.0
6X	7.8	6X	15.6
7X	8.5	7X	16.9
8X	9.0	8X	18.1
9X	9.5	9X	19.1
10X	10	10X	20
25X	14.0	25X	28.0
50X	17.0	50X	34.0
75X	18.8	75X	37.5
100X	20	100X	40

_____ $L = 20 \log q^{0.5}/q_0^{0.5} = 10 \log q/q_0$

where L is the sound power or energy

q is the sound power being measured

q_0 is the reference level

_____ Decibels increase as sound power increases (See Table 3.B.1).

_____ *h. Sound intensity calculations* (Ref. 36)

_____ Sound intensity defined as sound power per unit area

_____ Sound intensity = sound power/$4\pi r^2$

_____ Pictorial conceptualization of sound intensity (See Fig. 3.B.1.)

_____ 2. Noise-induced hearing loss (Ref. 33)

_____ *a.* Definition of noise—unwanted sound

_____ *b.* Threshold shifts (Ref. 33)

_____ (1) *Standard threshold shift*—A change (loss) in hearing of 10 dB or more occurs at 2000, 3000, and 4000 Hz.

Figure 3.b.1

___ (2) *Temporary threshold shift*—Hearing loss occurs, but hearing recovers within 14 hours.

___ (3) *Permanent threshold shift*—Hearing loss occurs permanently; factors for loss include

___ Prolonged exposure to sound levels in excess of 60 dB

___ Increased exposure to sound bursts

___ Inadequate breaks between sound exposures

___ 3. OSHA-permissible noise exposure limits (See Table 3.B.2.) (Ref. 1)

___ 4. Hearing protection

___ *a.* Applicability

___ Any employee exposed to an 8-hour TWA noise level of 85 dBA or greater

___ Any employee exposed to noise levels presented in Table 3.B.2

___ Any employee having experienced a standard threshold shift in hearing

TABLE 3.B.2 **OSHA-Permissible Noise Exposure Levels**

Duration Per Day (hours)	Sound Level (dBA)*
8	90
6	92
4	95
3	97
2	100
1.5	102
1	105
0.5	110
0.25 or less	115

* Sound level measured on A-scale of a standard sound level meter at slow response.

___ Any employee not having a baseline audiogram within the first six months of exposure

___ b. Noise abatement and control

___ Ensure proper preventive maintenance on equipment

___ Substitute less noisy equipment during upgrades and new installations (i.e., presses instead of hammers; belt drives instead of gears; larger, slower machines instead of smaller, faster machines)

___ Use less noisy processes (i.e., welding instead of riveting; pressing instead of rolling)

___ Reduce or dampen vibration of equipment

___ Use flexible mountings

___ Use sound-absorptive material on walls and ceilings

___ Direct sound away from the worker

___ Enclose or partially enclose machinery

___ Reduce pressure and flow velocities

___ Relocate machine to a room that is not occupied on a continuous basis

___ Provide sound-absorptive materials in the room
___ c. Types of hearing protection
   ___ Aural protectors (earplugs)
   ___ Superaural protectors (block external opening to ear)
   ___ Circumaural protectors (earmuffs)
   ___ Helmet protectors (enclose the whole head)
   ___ Use of only those devices tested/approved in accordance with ANSI standard S3.19

NOTES

NOTES

___ **C. Ergonomics** (Refs. 33 and 36)

___ 1. Definition and focus

    ___ Refers to the integration of worker, task, tools, and workstation to achieve a safe and comfortable working environment

    ___ Integrates many disciplines, such as anatomy, physiology, medicine, orthopedics, psychology, and sociology

    ___ Focuses on achieving an optimum relationship between the worker and the work environment

___ 2. Cumulative trauma disorders (CTDs)

    ___ *a.* Definition—Ergonomics-related problems resulting from repetitive motion or forceful motion such as vibration or mechanical compression or both

    ___ *b.* Common CTDs

        ___ (1) *Carpal tunnel syndrome*

            ___ Results from pressure being placed on the median nerve, which passes through the carpal tunnel of the wrist

            ___ Common with repetitive motion tasks, such as computer data entry, meat processing (deboning), assembly line work, machine operation control, letter sorting, and cashiering

        ___ (2) *De Quervain's disease*

            ___ An inflamation of the tendon sheath of the thumb resulting from excessive friction between two tendons and their common sheath

            ___ Caused by hand motions used in twisting and forceful gripping; common with tasks such as cutting, packing, and sewing

        ___ (3) *Hand-arm vibration syndrome*

            ___ Results from high-frequency vibrations which cause vascular tissues to

thicken, reducing or eliminating blood flow and causing nerve and vascular damage
___ Caused by use of vibrating tools
___ (4) *Raynaud's syndrome*
___ Results when damaged blood vessels cannot transport enough oxygen to skin and muscles in the hands
___ Caused from use of vibration tools
___ (5) *Tendinitis*
___ Results when the tendon of the wrist or shoulder becomes inflamed
___ Caused by overuse or unaccustomed use of wrist or shoulder; common among power press operators, welders, and assembly line workers
___ (6) *Tenosynovitis*
___ Results when there is an excessive secretion of synovial fluid from the synovial sheath surrounding the tendon
___ Caused from repetitive motion; can cause carpal tunnel syndrome (See above.)
___ (7) *Trigger finger*
___ Results when the tendon controlling the finger develops a bump or groove that prevents the tendon from sliding smoothly and locks the finger in a bent position
___ Caused from using tools with hard or sharp edges; is common among meat processors, assembly line workers, and carpenters
___ 3. *Ergonomics equations*
   ___ *a.* Work/rest cycle
     ___ $T_{rest} = (M_{max} - M/M_{rest} - M) \times 100$

where T_{rest} = Percentage of time that should be allowed for rest

M_{max} = Upper limit for metabolic cost for sustained work

M = Metabolic cost of the task

M_{rest} = Resting metabolism

___ Examples of calculations for time needed for rest (See Table 3.C.1.)

TABLE 3.C.1 **Examples of Work/Rest Cycles**

Task	Person Performing Task	M_{max} (kcal/hr)	M (kcal/hr)	M_{rest} (kcal/hr)	T_{rest} (% of day)
Pushing wheelbarrow	Fit young man	465	400	110	0
Pushing wheelbarrow	Less fit older man	350	400	110	17
Rock drilling	Fit young man	465	550	110	19
Rock drilling	Less fit older man	350	550	110	45

___ b. Equation for action limit for a load that over 75% of women and over 99% of men can lift

___ AL (kg) = $40(15/H)(1 - .004|V - 75|)(0.7 + 7.5/D)(1 - F/F_{max})$

___ AL (lb) = $90(6H)(1 - .01|V - 30|)(0.7 + 3/D)(1 - F/F_{max})$

where AL = Action limit

H = Horizontal hand location (in centimeters or inches)

V = Vertical location (in centimeters or inches)

D = Vertical travel distance (in centimeters or inches)

F = Average frequency of lift (in lifts/minute)

F_{max} = Maximum frequency of lift that can be maintained (in lifts/minute)

—— Maximum permissible limit (MPL) = 3 × AL

—— Examples of calculations for action limits and MPLs (See Table 3.C.2.)

TABLE 3.C.2 **Action Limit and Maximum Permissible Limit Calculations for Lifting**

Time Period	H (inches)	V (inches)	D (inches)	F (lifts/min)	F_{max} (lifts/min)	AL (lbs)	MPL (lbs)
1 hour	12	36	15	9	18	19	57
8 hours	12	36	15	9	15	15.2	45.6
1 hour	24	36	15	9	18	9.5	28.5
8 hours	24	36	15	9	15	7.6	22.8
1 hour	24	36	15	6	18	12.7	38
8 hours	24	36	15	6	15	11.4	34.2

—— 4. Workplace design guidelines

—— Avoid designs that incur static (isometric) muscle tension.

—— If static muscle tension cannot be avoided, ensure that the muscular load remains less than 15% of the maximal muscle force.

—— Design the work system to prevent overloading of the muscular system.

—— Forces necessary for dynamic activities should be kept to less than 30% of the maximal forces that the muscles are capable of generating; forces over 50% are acceptable only when kept to short duration.

—— Use postures for the limbs and body that provide the best lever arms for the muscles used.

—— For standing operations, cushion hard floors with soft floor mats.

—— Use foot pedals only for seated operations, as foot pedals tend to impose an imbalanced posture on the standing operator.

___ Maintain proper seating height, which allows the thighs to remain horizontal, the lower legs vertical, and the feet flat on the floor.

___ Use footrests, wrist rests, proper backrests, and other ergonomic features to prevent fatigue.

___ Design for allowing changes in posture.

___ Except for reach, design for the larger, taller operator, and make adjustments for the smaller, shorter operator.

NOTES

___ D. Industrial Ventilation

___ 1. Types of air contaminants
  ___ a. *Gas*
    ___ Has a very low density and viscosity
    ___ Can expand and contract greatly in response to changes in temperature and pressure
    ___ Easily diffuses into other gases
    ___ Readily and uniformly distributes itself throughout any container
    ___ Can be changed into the liquid or solid state only by the combined effect of increased pressure and decreased temperature (below the critical temperature)
  ___ b. *Vapor*—The gaseous phase of a substance that is a liquid at normal temperature and pressure
  ___ c. *Fume*
    ___ A gaslike emanation containing minute solid particles arising from the heating of a solid body, such as lead.
    ___ Finely divided solids produced by other methods of subdividing, such as chemical processing, combustion, explosion, or distillation.
    ___ The physical change is often accompanied by a chemical reaction such as oxidation.
    ___ Fumes are much finer than dusts, containing particles from 0.1 to 1 micron (μm) in size.
  ___ d. *Smoke*
    ___ Carbon or soot particles less than 0.1 micron in size
    ___ Results from incomplete combustion of carbonaceous materials such as coal or oil

___ *e. Mist*

___ Fine liquid droplets or particles measuring 40 to 500 microns

___ Generated by condensation from the gaseous state to the liquid state, or by breaking up a liquid into a dispersed state by splashing, foaming, or atomizing

___ 2. Principles for dilution ventilation design (Ref. 8)

___ Determine the amount of air needed to provide enough dilution to maintain the contaminant below its TLV. (See Table 3.D.1 for example of dilution air volumes needed for a specified room size for selected chemicals.)

___ Increase the quantity of dilution air to account for incomplete mixing of the air (termed K factor).

___ Choose locations for the air supply and exhaust outlets so that the air passes through the contaminated zone and does not transport contaminants into the breathing zone of workers.

___ Replace exhausted air by using a replacement air system.

TABLE 3.D.1   Example Dilution Air Volumes Needed for Selected Chemicals

Chemical Name	TLV (ppm)	Molecular Weight	Specific Gravity	Ventilation Rate (cfm)
Cyclohexane	300	84.2	0.78	621
Ethyl acetate	400	88.1	0.9	515
Ethyl benzene	100	106.2	0.9	1708
Isopropyl alcohol	400	60.09	0.785	658
Methyl ethyl ketone	200	72.1	0.8	1118

* Evaporation rate equals 1 pint per hour
** Assume incomplete mixing, with $K = 3$

___ Avoid reintroducing exhausted air by discharging the exhaust high above the roof line or by making sure that no window, outside air intakes, or other building openings are located near the exhaust outlet.

___ Sources for dilution ventilation include natural draft, thermal draft, and mechanical air movers.

___ 3. Dilution ventilation equation

$$Q = \frac{(403)(10^6(\text{SG})(\text{ER})(K)}{\text{MW}(C)}$$

where Q = ventilation rate in cubic feet per minute (cfm)

SG = specific gravity of contaminant as a liquid

ER = evaporation rate of the liquid in pints per minute

MW = molecular weight of the liquid

K = the adjusted factor for incomplete mixing

C = concentration of the gas or vapor, usually the TLV, expressed in ppm

___ 4. Local exhaust systems (Ref. 8)

___ a. Definitions (See Figs. 3.D.1 and 3.D.2.)

___ *Capture velocity*—Air velocity at any point in front of the hood or at the hood opening necessary to overcome opposing air currents and to capture the contaminated air at that point by causing it to flow into the hood

___ *Face velocity*—Air velocity at the hood opening

___ *Slot velocity*—Air velocity through the opening in a slotted hood

___ *Duct velocity*—Air velocity through the duct cross section

___ *Plenum velocity*—air velocity in the plenum

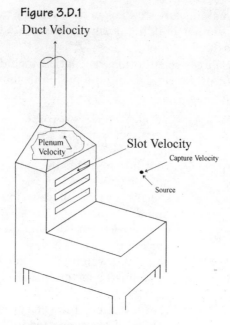

Figure 3.D.1
Duct Velocity

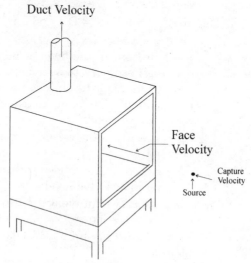

Figure 3.D.2
Duct Velocity

___ *b.* Design factors
    ___ (1) Sources of air motion
        ___ Thermal air currents, especially from heat-generating operations
        ___ Motion of machinery
        ___ Material motion
        ___ Movements of operator
        ___ Room air currents
    ___ (2) *Hood types/applications*
        ___ Canopy hood—Used for automated plating operations, degreasing operations, and hot processes
        ___ Open hood—Used for low- to medium-toxicity chemical-dispensing operations, flammable chemical-dispensing operations, lab operations
        ___ Slotted hood, side mounted—Used for brushing and/or grinding operations that produce low levels of dust
        ___ Slotted hood, downdraft—Used for operations involving fumes heavier than air and low-hazard bench operations
        ___ Enclosed hood—Used for dispensing of toxic chemicals, lab operations, extremely dusty operations, and high-agitation mixing operations
    ___ (3) Other factors
        ___ Minimum duct velocity based on transport velocity needed for specific contaminants
        ___ Branch duct size
        ___ Friction losses throughout the system
        ___ Airflow balancing
        ___ Fan type and pressure ratings
        ___ Variations in temperature and humidity

___ Air abatement devices or contaminant collectors

___ Types and locations of stacks

___ *c. Local exhaust equations*

___ $Q = VA$

where Q = air flow rate (cfm)

V = centerline velocity at the face of the hood

A = area of hood

___ Example required airflow rates for varying parameters for local exhaust (See Table 3.D.2.)

___ 5. Costs of ventilation equipment (Ref. 27)

___ *a.* Lab fume hood, installed—$1300

___ *b.* Glove box, fiberglass, bacteriological

___ Controlled atmosphere—$10,000

___ Radioisotope—$11,500

___ Carcinogenic—$13,000

TABLE 3.D.2   Example Required Air-Flow Rates for Varying Parameters for Local Exhaust

Hood Dimensions	Area of Hood (ft^2)	Air Velocity (fpm)	Required Air-flow Rate (cfm)
6-inch diameter	0.1963	2000	393
6-inch diameter	0.1963	4000	785
8-inch diameter	0.3491	4000	1396
12-inch diameter	0.7854	4000	3141

NOTES

NOTES

__ E. Personal Protective Equipment

__ 1. Hazards in the workplace that require PPE protection
  __ Impact
  __ Penetration
  __ Compression (rollover)
  __ Chemicals
  __ Heat
  __ Harmful dust
  __ Light radiation

__ 2. PPE requirements (OSHA, 29 CFR 1910)
  __ General requirements—1910.132
  __ Eye and face protection—1910.133
  __ Respiratory protection—1910.134
  __ Head protection—1910.135
  __ Foot protection—1910.136
  __ Electrical protective equipment—1910.137
  __ Hand protection—1910.138

__ 3. Eye and face protection
  __ Required when employees are exposed to potential hazards such as flying particles, molten metal, liquid chemicals, acid or caustic liquid, chemical gases or vapors, or potentially injurious light radiation.
  __ Devices must comply with ANSI Z87.1-1987, "American Standard for Occupational and Educational Eye and Face Protection."

__ 4. Respiratory protection
  __ a. Minimum requirements of a respiratory protection program
    __ Establish written operating procedures for selection and use of respirators.
    __ Select respirators on the basis of hazards to which the worker is exposed.
    __ Instruct and train the user in the proper use of respirators and their limitations.
    __ Provide for regular cleaning and disinfecting of respirators.

___ Provide a storage location for respirators that is convenient, clean, and sanitary.

___ Provide for the regular inspection of respirators.

___ Maintain appropriate surveillance of work-area conditions and the degree of employee exposure or stress.

___ Provide regular inspection and evaluation to determine the continued effectiveness of the program.

___ Ensure that persons assigned to tasks requiring the use of respirators have been determined to be physically able to perform the work and use the equipment.

___ Select respirators from among those approved by MSHA and/or OSHA

___ b. Respirator elements that should be inspected before and after each use

___ Facepiece

___ Head straps and head harness

___ Exhalation valve

___ Air-purifying apparatus

___ Breathing tube

___ Air supply system

___ c. Respirator Types

___ (1) *Chemical cartridge respirator*

___ Air-purifying respirators

___ Used for respiratory protection against specific chemicals of relatively low concentrations

___ Cannot be used in atmospheres containing less than 19.5% oxygen or that are IDLH

___ Available in half-face or full-face models

___ (2) *Particulate respirators*

___ Designed for protection against dusts, mists, fumes, and other particulates.

___ Atmosphere should be tested before use.

___ Cannot be used in atmospheres containing less than 19.5% oxygen or that are IDLH.

___ (3) *Gas masks*

___ Full-face masks to be used in an emergency

___ Can be powered or nonpowered

___ Not effective in atmospheres of less than 19.5% oxygen or in IDLH atmospheres

___ (4) *Supplied air respirators*

___ Also known as air-line respirators or atmospheric-supplying respirators.

___ Not approved for atmospheres containing less than 19.5% oxygen or in IDLH atmospheres because the air line could fail.

___ Full facepieces or hoods must be used with these devices.

___ (5) *Self-contained breathing apparatus (SCBA)*

___ Only respirator approved by NIOSH/MSHA for use in oxygen-deficient and IDLH atmospheres.

___ Also worn in emergencies and when unknown atmospheres must be entered.

___ Most widely used is open-circuit pressure-demand system, which maintains a positive pressure inside the facepiece at all times and provides the highest level of respiratory protection.

___ d. Defined canister (cartridge) colors for atmospheric contaminants

_____ Acid gases—White

_____ Hydrocyanic acid gas—White, with ½-inch green stripe completely around the canister near the bottom

_____ Chlorine gas—White, with ½-inch yellow stripe completely around the canister near the bottom

_____ Organic vapors—Black

_____ Ammonia gas—Green

_____ Acid gases and ammonia gas—Green with ½-inch white stripe completely around the canister near the bottom

_____ Carbon monoxide—Blue

_____ Acid gases and organic vapors—Yellow

_____ Hydrocyanic acid gas and chloropicrin vapors—Yellow, with ½-inch blue stripe completely around the canister near the bottom

_____ Acid gases, organic vapors, and ammonia gases—Brown

_____ Radioactive materials, excepting tritium and noble gases—Purple (Magenta)

_____ Particulates (in any combination with any of the aforementioned gases or vapors)

_____ Canister color for contaminant, as designated previously, with ½-inch gray stripe completely around the canister near the top

_____ All of the aforementioned atmospheric contaminants—Red, with ½-inch gray stripe completely around the canister near the top

_____ 5. Chemical protective clothing (CPC)—material compatibilities (Refs. 16 and 30)

_____ a. Butyl

_____ Compatible with moderate-to-strong acids, ammonia solutions, alcohols, inorganic salts, ketones, phenols, and aldehydes

___ Incompatible with alkanes, petroleum distillates, and solvents

___ Limited use with esters and ethers

___ *b.* Chlorinated polyethylene

___ Compatible with moderate-to-strong acids, bases, ammonium solutions

___ Limited use with ketones and solvents

___ *c.* Neoprene

___ Compatible with moderate acids, bases, ammonium solutions, alcohols, inorganic salts

___ Incompatible with petroleum distillates, esters

___ Limited use with ketones

___ *d.* Nitrile

___ Compatible with moderate-to-strong acids, bases, most ammonium solutions

___ Incompatible with phenols, ketones

___ Limited use with inorganic salts and aldehydes

___ *e.* Polycarbonates

___ Compatible with weak acids, ammonium solutions, alcohols, inorganic salts, phenols

___ Incompatible with aggressive petroleum distillates

___ Limited use with ketones

___ *f.* Polyvinyl chloride

___ Compatible with moderate-to-strong acids, bases, ammonium solutions, inorganic salts, alcohols, alkanes, and some aldehydes

___ Incompatible with petroleum distillates, ketones, concentrated solvents, phenols

___ *g.* Viton

___ Compatible with acids, bases, petroleum distillates, alcohols, inorganic salts

___ Incompatible with ketones

___ Limited use with aldehydes

___ 6. PPE during HAZMAT response

    ___ *a. Level A*

        ___ (1) Worn when highest level of protection is needed for respirator, skin, and eye protection

        ___ (2) Used in unknown atmospheres and known high-hazard atmospheres

        ___ (3) Protection includes the following:

           ___ SCBA

           ___ Fully encapsulating chemical-resistant suit

           ___ Chemical-resistant inner and outer gloves

           ___ Chemical-resistant safety boots

           ___ Hard hat, coveralls, long cotton underwear, cooling unit (as applicable)

    ___ *b. Level B*

        ___ (1) Worn when the highest level of respiratory protection is needed but a lesser level of skin and eye protection is required

        ___ (2) Typically used in oxygen-deficient and IDLH atmospheres, where substances do not present a skin hazard

        ___ (3) Protection includes the following:

           ___ SCBA

           ___ Chemical-resistant clothing, such as hooded shirt or jacket, one- or two-piece chemical splash-suit, disposable chemical coveralls, long-sleeved jacket, or equivalent clothing

           ___ Chemical-resistant inner and outer gloves

           ___ Chemical-resistant safety boots

           ___ Hard hat, coveralls, long cotton underwear, cooling unit (as applicable)

___ *c. Level C*

    ___ (1) Worn when the atmosphere contaminants are known and the concentrations have been measured and meet the criteria for use of an air-purifying respirator

    ___ (2) Worn when skin and eye exposure is unlikely

    ___ (3) Protection includes the following:

        ___ Full-facepiece air-purifying respirator

        ___ Chemical-resistant clothing

        ___ Chemical-resistant inner and outer gloves

        ___ Chemical-resistant safety boots

        ___ Escape mask (optional)

        ___ Hard hat, coveralls, long cotton underwear, cooling unit (as applicable)

___ *d. Level D*

    ___ (1) Worn when there is no respiratory hazard and when minimum skin and eye protection are required

    ___ (2) Primarily a work uniform

    ___ (3) Protection includes the following:

        ___ Coveralls

        ___ Safety glasses

        ___ Safety boots

___ 7. Costs of PPE (Ref. 27)

    ___ *a.* Respirators

        ___ Cartridge, ½ face—$13 to $20

        ___ Supplied air—$400

        ___ SCBA—$500 to $700

    ___ *b.* Chemical-resistant clothing

        ___ Lightweight chemical-resistant suit—$15 to 30

        ___ Acid King suit—$200 to $500

NOTES

___ F. Lighting

___ 1. General benefits of industrial lighting
    ___ Promotes higher quality work (fewer mistakes)
    ___ Reduces accidents
    ___ Increases production
    ___ Improves morale

___ 2. Minimum light levels needed for various tasks (in foot-candles) (Ref. 19)
    ___ *a.* Assembly operations
        ___ Rough easy-seeing—30
        ___ Rough difficult-seeing—50
        ___ Medium—100
        ___ Fine—500
        ___ Extra fine—1000
    ___ *b.* Automobile manufacturing
        ___ Frame assembly—50
        ___ Body parts manufacturing—70
        ___ Body and chassis assembly—100
        ___ Final assembly, finishing, and inspection—200
    ___ *c.* Canning and preserving
        ___ Labeling and cartoning—30
        ___ General handpacking—50
        ___ Can unscramblers—70
        ___ Cutting, pitting, final sorting, continuous belt canning, sink canning—100
    ___ *d.* Central facilities
        ___ Air conditioning and cable equipment, compressors, switch gears, gauge-area burner platforms, telephone equipment—10
        ___ Visitors gallery—20
    ___ *e.* Clothing manufacture
        ___ Receiving, storing, shipping, winding, measuring—30
        ___ Patternmaking, trimming—50
        ___ Marking—100

__ Cutting, pressing—300
__ Sewing, inspection—500

__ *f.* Electrical equipment manufacturing
__ Impregnating—50
__ Insulating, coil winding, testing—100

__ *g.* Foundries
__ Annealing furnaces—30
__ Cleaning—30
__ Pouring, sorting—50
__ Coke making—100
__ Moulding—100
__ Medium inspection—100
__ Fine inspection—500

__ *h.* Glass manufacturing
__ Mix and furnace room, glass-blowing machines—30
__ Grinding, cutting, slivering—50
__ Fine grinding, leveling, polishing—100

__ *i.* Inspection
__ Ordinary—50
__ Difficult—100
__ Highly difficult—200
__ Very difficult—500

__ *j.* Iron and steel manufacturing
__ Stock, hot top, checker-cellar calcining—10
__ Building, slag pits, stripping yard—20
__ Control platforms, mixer building, repairs—30
__ Rolling mills—30
__ Motor room and machine room—30
__ Shearing and tin plate—50
__ Inspection—100

__ *k.* Machine shops
__ Rough bench—50
__ Medium bench, rough grinding, buffing—100

___ *l.* Materials handling
  ___ Loading, trucking—20
  ___ Picking stocks, classifying—30
  ___ Wrapping, packing, labeling—50
___ *m.* Meat packing
  ___ Slaughtering—30
  ___ Cleaning, canning—100
___ *n.* Offices
  ___ Corridors, stairways—20
  ___ Reading, transcribing—70
  ___ Regular office work—100
  ___ Accounting, auditing, tabulating, business-machine operation, rough drafting—150
  ___ Cartography, designing, detailed drafting—200
___ *o.* Paint Shop
  ___ Spraying, rubbing, hand art, stencil—50
  ___ Fine hand painting and finishing—100
___ *p.* Paper manufacturing
  ___ Beaters, grinding—30
  ___ Finishing, cutting—50
  ___ Hand counting—100
  ___ Paper machine reel, inspection—100
  ___ Rewinder—150
___ *q.* Printing
  ___ Photo engraving, etching, blocking—50
  ___ Presses—70
  ___ Composing room—100
  ___ Proofreading—150
  ___ Color inspecting—200
___ *r.* Rubber goods and tire manufacturing
  ___ Plasticating, milling—30
  ___ Fabric cutting, hose looms, molding—50
  ___ Building, wrapping, curing—70
___ *s.* Sheet metal works
  ___ General—50

___ Tin plate inspection, galvanized scribing—200

___ *t.* Silk and synthetics

___ Soaking, tinting, conditioning—30

___ Winding, slashing, light thread—50

___ Winding, slashing, dark thread—200

___ *u.* Textile Mills

___ Picking, carding, roving, spinning—50

___ Beaming and slashing—150

___ Drawing—200

___ *v.* Welding, general—50

___ *w.* Warehousing

___ Inactive—5

___ Active, rough bulky—10

___ Active, medium—20

___ Active, fine—50

___ *x.* Woodworking

___ Rough sawing and bench work—30

___ Sizing, planing, gluing, sanding, veneering, medium quality bench work—50

___ Fine bench work, fine sanding, finishing—100

NOTES

NOTES

PART 4

Safety Management

NOTES

__ A. Accident/Loss Prevention and Control

___ 1. Causes of accidents/illnesses
    ___ *a.* Human factors
        ___ Training/skills not adequate
        ___ Procedure not followed
        ___ Unsafe practices used
        ___ Deviation from safety rules
        ___ Hazard not recognized
    ___ *b.* Situational Factors
        ___ Equipment design defects
        ___ Substandard construction
        ___ Improper storage of hazardous materials/equipment
        ___ Inadequate facility layout
    ___ *c.* Environmental Factors
        ___ Physical factors such as noise, lighting, or vibration
        ___ Chemical exposures to dusts, fumes, gases, vapors, or mists
        ___ Biological factors such as sensitivities, age, sex, strength, or conditioning
        ___ Ergonomic factors such as repetitive motion, lifting, and workstation design
___ 2. Examples of accidents/illnesses and potential causes (See Table 4.A.1.)
___ 3. Classes of accidents
    ___ Class 1—Cases involving lost workdays, including days away from work or days of restricted work activity
    ___ Class 2—Medical treatment cases requiring the attention of a physician outside the plant
    ___ Class 3—Medical treatment cases requiring only first aid or local dispensary treatment and resulting in property damage of less than $100 or loss of less than eight hours in work time
    ___ Class 4—Accidents that either cause no injury or cause minor injury not requiring the attention of a physician, and that result in property damage

of $100 or more or loss of eight or more employee-hours

TABLE 4.A.1   Examples of Accidents/Illnesses and Potential Causes

Accident/Illness	Potential Causes
Occupational illness as a result of exposure and/or overexposure to a chemical(s)	1. Failure to use chemical-protective equipment or improper use of the equipment 2. Inadequate chemical-protective equipment 3. Inadequate training on chemical hazards 4. Inadequate procedures for safe handling and storage of chemicals
Slipping or tripping	1. Slipping/tripping hazard not identified 2. Spills not cleaned up 3. Poor housekeeping 4. Employee not watching when walking/working
Electrical shock	1. Use of ungrounded or uninsulated tools 2. Failure to use protective gloves and boots 3. Improper use of electrical equipment 4. Improper training/skills to work with electrical equipment 5. Inadequate procedures for safe use of electrical equipment 6. Deviation from procedures for use of electrical equipment
Falls	1. Failure to use lanyard or other fall-protection device 2. Defective fall-protection device and/or device not inspected before use 3. Employee not watching when walking/working

TABLE 4.A.1 Examples of Accidents/Illnesses and Potential Causes (*Continued*)

Accident/Illness	Potential Causes
Cuts, abrasions, punctures, lacerations	1. Failure to use protective clothing 2. Improper use of mechanical equipment 3. Removing or disabling guards 4. Inadequate equipment design 5. Hazard not identified 6. Inadequate procedures or procedures not followed
Bruise, contusion	1. Improper use of equipment 2. Inadequate equipment design 3. Failure to use protective equipment
Concussions	1. Heavy objects put away improperly 2. Failure to wear protective equipment 3. Unsafe area not properly marked
Strains or sprains	1. Poorly designed workstation. 2. Improper lifting or other movement 3. Inadequate physical conditioning for task 4. No work rest or inadequate work rest
Burns or scalds	1. Burn hazard not identified 2. Failure to use proper protective equipment 3. Inadequate protective equipment 4. Faulty equipment
Amputation	1. Improper use/operation of mechanical equipment 2. Failure to use proper protective equipment 3. Improper training/skills for task 4. Removing or disabling guards

___ 4. OSHA recordkeeping requirements (Ref. 32)

    ___ *a.* OSHA No. 200 (form)

        ___ Used for recording and classifying occupational injuries and illnesses

        ___ Used for noting the extent of each case

        ___ Provides information on the employee and the nature and extent of the injury or illness

    ___ *b.* OSHA No. 101 (form)

        ___ Used to record additional information for every injury or illness entered on the OSHA No. 200 form

        ___ Alternate reports—including worker's compensation, insurance, or other reports—may be used in lieu of this one if the required information is covered

    ___ *c.* Annual summary of occupational injuries and illnesses

        ___ Part of OSHA No. 200 form

        ___ Summary includes total of injuries and illnesses for the calendar year

    ___ *d.* Industry segments subject to recordkeeping requirements

        ___ Agriculture, forestry and fishing

        ___ Oil and gas extraction

        ___ Construction

        ___ Manufacturing

        ___ Transportation and public utilities

        ___ Wholesale trade

        ___ Building materials and garden supplies

        ___ General merchandise and food stores

        ___ Hotels and other lodging places

        ___ Repair services

        ___ Amusement and recreation services

        ___ Health services

    ___ *e.* Categories of recordable cases

        ___ (1) *Fatalities*—All must be recorded regardless of the time between the injury/illness and the death

___ (2) *Lost workday cases*—Occur when the injured or ill employee experiences either days away from work, days of restricted work activity, or both

___ (3) Other recordable cases—Typically less serious than the above types of cases but satisfy the criteria for recordability, including the following:

    ___ Those requiring medical treatment (does not include first-aid treatment only)

    ___ Those involving loss of consciousness

    ___ Those involving restriction of work or motion

    ___ Those requiring transfer to another job

___ *f.* Incident rates

    ___ Incidence rate—No. of injuries and illnesses times (×) 200,000, divided by (÷) total hours worked by all employees for period covered

    ___ Average lost workdays per total lost workday cases—Total lost workdays divided by (÷) total lost workday cases

    ___ Average days away from work—Total days away from work divided by (÷) the total cases involving days away from work

___ 5. Worker's Compensation

    ___ *a.* Objectives of Worker's Compensation

        ___ Provide adequate, equitable, prompt, and sure income and medical benefits to work-related accident victims, or income benefits to their dependents, regardless of fault

        ___ Provide a single remedy and reduce court delays, costs, and workloads arising out of personal litigation

        ___ Relieve public and private charities of financial drains—incident to uncompensated industrial accidents

   ___ Eliminate payment of fees to lawyers and witnesses as well as time-consuming trials and appeals

   ___ Encourage maximum employer interest in safety and rehabilitation through an appropriate experience-rating mechanism

   ___ Promote frank study of causes of accidents (rather than concealment of fault), reducing preventable accidents and human suffering

  ___ *b.* Covered injuries

   ___ Personal injury caused by accidents arising out of and in the course of employment

   ___ Occupational diseases arising out of and in the course of employment

   ___ Occupational hearing loss

   ___ Black lung disease

   ___ Work-related impairment

  ___ *c.* Types of Benefits

   ___ Income replacement

   ___ Medical benefits

   ___ Rehabilitation (medical and vocational)

___ 6. Safety management for accident prevention

 ___ Includes integration of behavior and personal approaches

 ___ Avoids placing blame on individual

 ___ Focuses on the design of the process or activity in order to reduce accidents

 ___ Uses observations of behavior and the sharing of these observations with workers

 ___ Involves worker participation in defining and solving safety problems

 ___ Utilizes recognition and incentives for workers

___ 7. Top 10 OSHA violations in fiscal year 1997 (Oct. 1–Sept. 30) (Ref. 23)

___ Hazard Communication (29 CFR 1910.1200)—23%
___ Scaffolding: Construction (29 CFR 1926.451)—14%
___ Lockout/Tagout (29 CFR 1910.147)—10%
___ Fall Protection (29 CFR 1926.501)—10%
___ Electrical: Wiring Methods, Components and Equipment for General Use (29 CFR 1910.305)—9%
___ Mechanical Power Transmission (29 CFR 1910.219)—8%
___ Machine Guarding: General Requirements (29 CFR 1910.212)—8%
___ Electrical: General Requirements (29 CFR 1910.303)—6%
___ Personal Protective Equipment (1910.132)—6%
___ Respirators (1910.134)—6%

NOTES

__ B. Electrical Safety

__ 1. Electric current
    __ *a.* General
        __ Is produced when free electrons in a wire or other conductive material are moved in the same direction.
        __ Electric current is a function of electrical potential between two points and the resistance between them.
        __ Ohm's law (see Sec. 3.B.3 below).
    __ *b.* Resistance
        __ If there is more than one path between two points that have differing electrical energy levels, the current will travel through the path of least resistance.
        __ Resistance of a wire is directly proportional to its length.
        __ Resistance of a wire is inversely proportional to its cross section.
        __ Conductance is equal to 1/resistance.
__ 2. Electrical circuits—definitions
    __ General definition—Closed circuits made up of a power source, electric conductors (wire), and a load
    __ *Series loads*—Total circuit resistance is the sum of the resistances of each individual load
    __ *Power*—The amount of work that can be done by a load in some standard amount of time
    __ *Parallel circuit*—A circuit in which there are one or more points where the current divides and follows different paths before recombining to flow back to the power source
__ 3. General equations for electrical circuits (See Fig. 4.B.1.)
    __ *a. Ohm's law and variations of Ohm's law*
        __ $I = V/R$
        where I is in amperes
        V is in volts
        R is in ohms

Figure 4.B.1

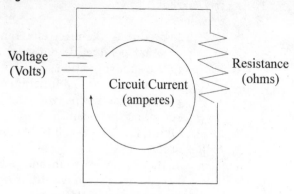

Electrical Circuit

___ Circuit Current $(I) = V/R$
___ Voltage $(V) = IR$
___ Resistance $(R) = V/I$
___ b. Other equations
___ Total resistance in a series load $(R_{TOT}) = R_1 + R_2 + R_3$ (See Fig. 4.B.2.)

Figure 4.B.2

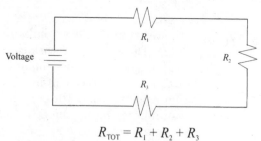

$$R_{TOT} = R_1 + R_2 + R_3$$

___ Power $(P) = I^2R$, where P is in watts
___ Power $(P) = VI$
___ Total resistance in a parallel circuit $(1/R_{TOT}) = 1/R_1 + 1/R_2 + 1/R_3$ (see Fig. 4.B.3)

Figure 4.B.3

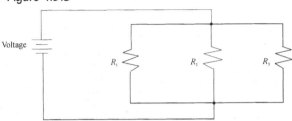

$$1/R_{\text{TOT}} = 1/R_1 + 1/R_2 + 1/R_3$$

___ 4. OSHA-defined hazard locations

    ___ a. *Class I, Division 1*

        ___ (1) Definitions

            ___ A location in which hazardous concentrations of flammable gases or vapors may exist under normal operating conditions

            ___ A location in which hazardous concentrations of such gases or vapors may exist frequently because of repair or maintenance operations or because of leakage

            ___ A location in which breakdown or faulty operation of equipment or processes might release hazardous concentrations of flammable gases or vapors and might also cause simultaneous failure of electric equipment

        ___ (2) Examples

            ___ Flammable liquid and gas transfer stations

            ___ Interiors of spray booths where flammable solvents are used

            ___ Areas in the vicinity of spraying and painting operations where volatile flammable solvents are used

___ Locations containing open tanks or vats of volatile flammable liquids

___ b. *Class I, Division 2*

    ___ (1) Definitions

        ___ A location in which volatile flammable liquids or flammable gases are handled, processed, or used in closed containers or closed systems and in which the liquids vaporize

        ___ A location in which gases or vapors are normally prevented by positive mechanical ventilation but could become hazardous if the ventilation system fails

        ___ A location adjacent to a Class I, Division 1 location to which hazardous concentrations of gases or vapors might occasionally be communicated

    ___ (2) Examples

        ___ Closed-loop still area

        ___ Ventilated flammable liquid–dispensing area

        ___ A location adjacent to a spray booth that is not protected by positive pressure ventilation or other safeguard against ventilation failure

___ c. *Class II, Division 1*

    ___ (1) Definitions

        ___ A location in which combustible dust is or may be in suspension in the air under normal operating conditions in quantities sufficient to produce explosive or ignitible mixtures

        ___ A location where mechanical failure or abnormal operation of machinery or equipment might cause such explosive or ignitible mixtures to be produced and might also provide a source of ignition through simultaneous failure of electric equipment, operation of protection devices, or through other causes

___ A location in which combustible dusts of an electrically conductive nature may be present

___ (2) Examples

___ Grain silos

___ Fueling stations

___ Chemical (solids) silos

___ d. *Class II, Division 2*

___ (1) Definitions

___ A location in which combustible dust will not normally be in suspension in the air in quantities sufficient to produce explosive or ignitible mixtures, and dust accumulations are normally insufficient to interfere with the normal operation of electrical equipment or other apparatus.

___ Dust may be in processing equipment, and resulting dust accumulations may be ignitible by abnormal operation or failure of electrical equipment in other apparatus.

___ (2) Examples

___ Warehouse areas for chemical bags of solids

___ Areas which house closed pneumatic conveyors of solid (dusty) materials and/or chemicals

___ e. *Class III, Division 1*

___ (1) Definition—A location in which easily ignitible fibers or materials producing combustible flyings are handled, manufactured, or used

    ___ (2) Example—Areas which house textile-manufacturing operations

___ *f. Class III, Division 2*

    ___ (1) Definition—A location in which easily ignitible fibers are stored or handled, except in process of manufacture

    ___ (2) Example—Areas of textile storage and/or packaging

___ 5. Control of hazardous energy—lockout/tagout procedures

    ___ *a.* Applications

        ___ Lockout of energy-isolating device is required when an employee performs any servicing or maintenance on a machine or equipment where the unexpected energizing, start up, or release of stored energy could occur and cause injury.

        ___ If energy-isolating device cannot be locked out, a tagout system must be used.

        ___ Equipment which is replaced, undergoes major repair, renovation, or modification, and whenever there is installation of new equipment, the equipment must be designed to accept a lockout device.

        ___ Hazardous energy sources could include electrical energy, energy from compressed air or gases, or stored energy from pressurized liquids in piping.

    ___ *b. Lockout/tagout requirements checklist* (See Table 4.B.1.)

TABLE 4.B.1   **Lockout/Tagout Requirements Checklist**

Energy Control Procedures

- Must be documented
- Must specify steps for shutting down, isolating, blocking and securing machines or equipment to control hazardous energy
- Must specify steps for the placement, removal, and transfer of lockout and/or tagout devices and the responsibility for them
- Must specify specific requirements for testing a machine or equipment to determine and verify the effectiveness of lockout/ tagout devices and other energy control measures

Periodic Inspections of the Energy Control Procedures

- Must be conducted at least annually
- Must be performed by an authorized employee who is not using the energy control procedure being inspected
- Must be designed to identify and correct any deviation or inadequacies
- Must include a review between the inspector and authorized employee(s) to ensure the employee(s) understand responsibilities under the procedure
- Must be certified by employer

Employee Training

- Must ensure that the purpose and function of the energy control program are understood by employees and that the knowledge and skills required for the safe application, usage, and removal of energy controls are acquired by employees
- Authorized employees—Those who isolate and work on equipment that is a source for hazardous energy; must receive training in recognition of applicable hazardous energy sources, the type and magnitude of energy available in the workplace, and the methods and means necessary for energy isolation and control
- Affected employees—Those who work at the equipment but do not perform the energy isolation and maintenance tasks; must be instructed in the purpose and use of the energy control procedure
- Other employees—Those who work in the area or might be in the area; shall be instructed about the energy control procedure and about the prohibition relating to attempts to restart or reenergize machines or equipment that is locked out or tagged out
- Training in tagout systems (where used) must provide and include the following information:
  —Tags are essentially warning devices affixed to energy-isolating devices and do not provide physical restraint on those devices provided with a lock.

TABLE 4.B.1 **Lockout/Tagout Requirements Checklist**
(*Continued*)

Employee Training (Continued)

—When a tag is attached to an energy-isolating means, it is not to be removed without authorization of the authorized person responsible for it, and it is never to be bypassed, ignored, or otherwise defeated.

—Tags must be legible and understandable by all authorized employees, affected employees, and other employees who may be in the area.

—Tags and their means of attachment must be made of materials which will withstand the environmental conditions in the workplace.

—Tags may evoke a false sense of security; their meaning must be clearly understood.

—Tags must be securely attached to energy-isolating devices so that they cannot be inadvertently or accidentally detached during use.

Employee Retraining

• Conducted for authorized and affected employees when
  —There is a change in their job assignment
  —There is a change in machines, equipment, or processes that present new hazards
  —There is a change in energy-control procedures
• Conducted when a periodic inspection reveals that there are deviations from or inadequacies in an employee's knowledge or use of the energy control procedures
• Must reestablish employee proficiency
• Must introduce new or revised control methods and procedures

Employer Certification

• Attests that employee training has been accomplished
• Attests that employee training is up to date
• Includes employee's name and dates of training

Lockout and Tagout Devices

• Must be provided by the employer
• Must be singularly identified
• Must be used only for controlling energy and not for other purposes
• Must be standardized within the facility through color or shape or size

TABLE 4.B.1   **Lockout/Tagout Requirements Checklist**
(*Continued*)

Lockout and Tagout Devices (Continued)

- Must be capable of withstanding the environment for the maximum period of time exposure is expected
- Lockout devices must
  - Be substantial enough to prevent removal without the use of excessive force or unusual techniques such as the use of bolt cutters or metal-cutting tools
  - Be attached in a manner that will hold the energy-isolating device in a safe or off position
- Tagout devices must
  - Have standardized print and format
  - Be substantial enough to prevent inadvertent or accidental removal
  - Warn against hazardous conditions if the machine or equipment is energized
  - Include warnings such as Do Not Operate, Do Not Open, or Do Not Energize
  - Be affixed in a manner that clearly indicates that the operation or movement of energy-isolating devices from the safe or off position is prohibited

NOTES

___ C. Machine Safety

___ 1. Hand and power tools
   ___ *a.* Tool types
      ___ (1) Hand tools—Nonpowered tools such as hammers, screwdrivers, crowbars, and wrenches
      ___ (2) Power tools
         ___ *Electrical tools*—Electric-powered tools such as saws, drills, and grinders
         ___ *Pneumatic tools*—Powered by compressed air; include tools such as nailers, staplers, hammers, and sanders
         ___ *Liquid fuel tools*—Typically powered by gasoline; include tools such as mowers, trimmers, and generators
         ___ *Powder-actuated tools*—Function like a loaded gun; include tools such as nailers and staplers
         ___ *Hydraulic power tools*—Powered by a fluid; include hydraulic jacks and lifts
   ___ *b.* Tool safety practices
      ___ (1) General
         ___ Ensure tools are maintained and are in good condition.
         ___ Ensure that the tool used is right for the job.
         ___ Inspect tools before starting the job, and do not use if damaged.
         ___ Review and follow manufacturer's instructions for tool operation.
         ___ Ensure proper protective equipment is used.
      ___ (2) Electric tools
         ___ Tool must have three-wire cord with ground and be grounded; or

——— Tool must be double insulated (most typical); or

——— Tool must be powered by low-voltage isolation transformer.

——— (3) Pneumatic tools

——— Tools must have positive locking devices for attachments, safety clips, or retainers to prevent attachments from being unintentionally shot from the tool barrel.

——— Hearing protection may be required when using these tools.

——— (4) Liquid fuel tools

——— There must be proper ventilation when used in closed areas to dissipate fumes; or

——— Respiratory protection must be used.

——— Fire extinguishers must be kept in the area.

——— (5) Powder-actuated tools

——— Muzzle end of the tool should have protective guard or shield.

——— Ear, eye, and face protection must be used, as required.

——— Defective tools must be tagged and removed from service immediately.

——— (6) Hydraulic power tools

——— Fluid must be an approved fire-resistant fluid.

——— Fluid must retain its operating characteristics at the most extreme temperatures to which it will be exposed.

——— The base of the tool must rest on a firm and level surface.

——— 2. Machine guarding (Ref. 34)

——— a. Guard types

——— (1) *Fixed guards*

___ Enclose a machine or otherwise pro-
tect the operator from moving parts
and other hazards

___ Prevent accidents to fingers, hands,
and other body parts

___ Prevent loose clothing from being
caught by moving parts

___ (2) *Interlocked guards*

___ Designed to prevent operation of a
machine if the guard is raised for
purposes of machine setup, adjust-
ment, or maintenance

___ Allow for access to machine for
removing jams without removing
fixed guards

___ (3) *Adjustable guards*

___ Designed to allow for a variety of production operations while maintaining hazard protection

___ Can be adjusted to admit varying
sizes of stock

___ (4) *Self-adjusting guards*

___ Designed for a tool or machine

___ Float or otherwise self-adjust to
allow for movement of materials
while safeguarding the person operating the machine

___ *b.* Safety devices and controls

___ (1) *Moveable barriers*

___ Designed to allow materials to be
inserted or removed

___ Interlocked with machine so that
machine is not operable when barrier is opened

___ (2) *Automatic feed systems*

___ Designed to prevent contact of operator with the hazardous parts of the
machine

___ Machine shut down critical if jamming occurs.

___ (3) *Presence-sensing devices*

___ Mats, photoelectric sensors or other devices that sense the presence of a person or body in the hazard zone

___ Automatically shut off machine operation when presence is sensed

___ (4) *Emergency stop control*

___ Stops the operation of a machine immediately

___ Typically a button or lever

___ Used for emergency situations

___ (5) *Hand control devices*

___ Used to ensure that operators' hands are not at the point of operation

___ Typically pull-out devices, restraint devices, or two-hand control devices

___ (6) *Low-energy and inch controls*

___ Used to reduce the speed (and hazards) of equipment

___ Used during setup, cleaning, and maintenance activities

NOTES

NOTES

___ D. Process and System Safety

___ 1. Process safety management (PSM)
   ___ *a. PSM standard*
      ___ Delineated in 29 CFR 1910.119
      ___ Sets for requirements for an organized approach to PSM
   ___ *b. PSM checklist* (See Table 4.D.1.)

TABLE 4.D.1 PSM Checklist

Process Safety Information

- Hazard information for each highly hazardous chemical
- Information pertaining to the technology of the process
- Information pertaining to equipment used in the process

Employee Involvement

- Involvement in developing and implementing PSM elements
- Involvement in process hazard analysis

Process Hazard Analysis

- Must address the following:
  —Hazards of the process
  —Identification of any previous incident that has a potential for catastrophic consequences in the workplace
  —Engineering and administrative control applicable to hazardous workplace elements and their interrelationships
  —Consequences of failure of engineering and administrative controls
  —Facility sitings
  —Human factors
  —Qualitative evaluation of a range of the possible safety and health effects on employees in the workplace if there is a failure of controls
- Process hazard analysis methodologies include:
  —"What if"—Used for relatively uncomplicated processes; effects of component failures and procedural or other errors on various aspects of the process are analyzed
  —Checklist—Used for more complex processes; a checklist is developed, and aspects of the process are assigned to committee members who have experience or skill to evaluate those aspects
  —What if/checklist—Combines the what-if and checklist processes to provide a comprehensive approach to process evaluation
  —Hazard and operability study (HAZOP)—Used to systematically investigate each element of a system for all potential deviations from design conditions
  —Failure mode and effects analysis (FMEA)—Used to methodically study component failure, including potential mode of failure, consequence of failure, probability of failure, and compensating provisions

TABLE 4.D.1   **PSM Checklist** (*Continued*)

Process Hazard Analysis (Continued)

—Fault tree analysis—Used to model all undesirable outcomes from a specific event

Operating Procedures

- Written to provide clear instructions for safely conducting activities involved in each process
- Include steps for each operating phase, including start-up, normal operations, temporary operations, emergency operations, and shutdown
- Describe safe operating limits and consequences of deviation from these limits
- Provide safety and health considerations, such as properties and hazards of chemicals, safety systems, and control measures, should physical contact or airborne exposure occur

Contractors

- Employer responsibilities
  - Obtaining and evaluating information regarding the contract employer's safety performance and programs
  - Informing contract employers of the known potential fire, explosion, or toxic release hazards related to the contractor's work and the process
  - Explaining to contract employers the applicable provisions of the emergency action plan
  - Developing and implementing safe work practices consistent with PSM standards to control the entrance, presence, and exit of contract employers and contract employees in covered process areas
  - Evaluating periodically the performance of contract employers in fulfilling their obligations under the PSM standard
  - Maintaining a contract employee injury and illness log related to the contractor's work in the process areas
- Contractor employer responsibilities
  - Ensuring that the contract employees are trained in the work practices necessary to perform their jobs safely
  - Ensuring that the contract employees are instructed in the known potential fire, explosion, or toxic release hazard related to their jobs and the process, and in the applicable provisions of the emergency action plan
  - Documenting that each contract employee has received and understood the standard training required by preparing a record that contains the identity of the contract employee, the date of training, and the means used to verify that the employee understood the training
  - Ensuring that each contract employee follows the safety rules of the facility, including the required safe work practices required in the operating procedures
  - Advising the employer of any unique hazards presented by the contract employer's work

TABLE 4.D.1   **PSM Checklist** (*Continued*)

Pre-start-up Safety Review

- Required for new or modified processes
- Includes confirmation that equipment meets design specifications and that safety, operating, maintenance, and emergency procedures are in place
- Ensures a process hazard analysis has been performed for new facilities

Mechanical Integrity of Equipment

- Includes pressure vessels, storage tanks, piping systems, relief and vent systems and devices, and sensors, alarms, and interlocks
- Requires procedures for maintaining the ongoing integrity of process equipment
- Requires inspections and tests on the process equipment, using the following criteria:
  —Procedures for inspections and tests must follow recognized and generally accepted good engineering practice.
  —Specification of frequency of inspections and tests must be consistent with manufacturer's recommendation.
  —Testing documentation must include the date of the inspection or test, the name of the person who performed the inspection or test, the serial number or other identifier of the inspected/tested equipment, a description of the inspection or test performed, and the results of the inspection or test.
- Correct equipment deficiencies before further use unless other steps are taken to ensure safe operation
- For new operations, ensure equipment is suitable for the process application for which it will be used and that appropriate field checks and inspections are performed to ensure that equipment is installed properly

Hot Work Permits

- Issued by employers for hot work operations on or near process
- Documents that fire prevention and protection requirements are met
- Kept on file until work is complete
- Filed at facility for a period of time designated by the safety professional (good management practice)

Management of Change

- Applies to changes in process chemicals, technology, equipment, and procedures
- Written procedures required which address the technical basis for the proposed change, the impact of the change on employee safety and health, modifications to operating procedures, necessary time period for the change to be effected, and the authorization requirement for the change

TABLE 4.D.1   PSM Checklist (*Continued*)

Management of Change (Continued)

- Employee communication and training required before start-up
- Update to information, procedures, and other documents required before start-up

Incident Investigation

- Applies to incidents that resulted in, or reasonably could have resulted in, a catastrophic release of a highly hazardous chemical in the workplace
- Initiated as promptly as possible, but within 48 hours
- Investigation team established with at least one person knowledgeable about the process
- Report required that includes date and time of the incident, the date the investigation began, a description of the incident, factors contributing to the incident, and recommendations resulting from the investigation
- Investigation report saved for 5 years

Emergency Planning and Response

- Required for the entire facility
- Includes procedures for handling small releases

Compliance Audits

- Required every three years to ensure compliance with provisions of PSM
- Includes a written-findings report and documentation of corrected deficiencies
- Two most recent reports kept on file

Trade Secrets

- All information, including trade secrets, is to be used for carrying out elements of PSM
- Confidentiality agreements for nondisclosure of trade secrets allowed

____ 2. System reliability

    ____ *a. Series systems*

        ____ System is made up of several components in series.

        ____ Components function independently from one another.

        ____ If one component fails, the system fails.

        ____ Reliability of the system is calculated by multiplying the reliabilities of the individual components (See Table 4.D.2).

TABLE 4.D.2 **Reliability of Series Systems: Reliability Rate**

Component A	Component B	Component C	Component D	Series System*
0.98	0.98	0.98	0.98	0.92
0.98	0.98	0.95	0.90	0.82
0.98	0.97	0.95	0.90	0.72
0.95	0.92	0.85	0.85	0.63
0.92	0.92	0.88	0.79	0.59

* Series system reliability rate is calculated by multiplying the reliabilities of the individual components.

___ b. *Parallel systems*
    ___ Components are operated in parallel.
    ___ Failure occurs only if all components fail.
    ___ Reliability of the system is calculated using the products of unreliability (or failure) as follows:
    ___ $P_s = 1 - P_f$
    where P_s = Probability of success, and
            P_f = Probability of failure
    ___ Probability of failure is calculated as follows:
    ___ $P_f = 1 - P_s$
    ___ Example rates of reliability for parallel systems (See Table 4.D.3.)

TABLE 4.D.3 **Reliability of Parallel Systems**

Reliability of Component A	Reliability of Component B	Probability of Failure	Reliability of Parallel System
0.9	0.9	0.01	0.99
0.9	0.8	0.02	0.98
0.85	0.85	0.0225	0.9775
0.85	0.8	0.03	0.97
0.9	0.75	0.025	0.975
0.75	0.75	0.0625	0.975

NOTES

___ E. Confined Space Safety

___ 1. Definitions of confined spaces

    ___ (a) *OSHA definition of confined space*—A confined space is one that has limited or restricted means of entry or exit, is large enough for an employee to enter and perform work, and is not designed for continuous occupancy by the employee. Examples include the following:

        ___ Underground vaults

        ___ Tanks

        ___ Storage bins

        ___ Pits

        ___ Diked areas

        ___ Vessels

        ___ Silos

    ___ (b) *Permit-required confined space*—One that meets the OSHA definition of a confined space and has one or more of the following characteristics:

        ___ The confined space contains or has the potential to contain a hazardous atmosphere.

        ___ The confined space contains a material that has the potential for engulfing an entrant.

        ___ The confined space has an internal configuration that might cause an entrant to be trapped or asphyxiated by inwardly converging walls or by a floor that slopes downward and tapers to a smaller cross section.

        ___ The confined space contains any other recognized serious safety or health hazards.

___ 2. *Confined space entry permit checklist* (See Table 4.E.1.)

TABLE 4.E.1   **Confined Space Permit Checklist**

Test Data

- Initial atmospheric test results for oxygen, explosives, and toxics
- Secondary test results after ventilation of space
- Test results of periodic monitoring of space
- Results of continuous monitoring for known contaminant
- Time(s) and date of test(s)
- Name or initials of person performing the test(s)

Equipment

- Monitoring equipment
  - O_2 meter
  - LEL meter
  - CO meter
  - Other: list
  - Calibration of Monitoring Equipment
    - Date and time
    - Name or initials of person performing the calibration
- Personal protective equipment
  - SCBA
  - Air line
  - Cartridge respirator
  - Dust mask
  - Encapsulating suit
  - Chemical suit
  - Gloves
  - Boots
  - Other: list
- Safety equipment
  - Safety harness
  - Lifeline
  - Hoisting equipment
  - Nonsparking tools
  - Explosion-proof lighting
  - Fire extinguishers
  - Other: list
- Communication Equipment
  - Two-way radio
  - Bullhorn
  - Pager
  - Evacuation alarms
  - Other: list

TABLE 4.E.1   **Confined Space Permit Checklist** (*Continued*)

Entry Information

- Facility name
- Permit space name or number
- Work to be performed
- Entry procedures
- Control measures
  —Secure/isolate area
  —Lockout/tagout of equipment
  —Ventilate or purge space
  —Acquire hot work permit
  —Other: list

Personnel Information

- Name and signature of supervisor who authorizes entry
- Name(s) of authorized entrant(s)
- Names of attendants
- Name of authorized entry supervisor
- Names of standby personnel
- Confined space training information
- Names and telephone numbers for on-site and off-site emergency services

---

Other

- Emergency plans and procedures
- Job-specific information
- Comments section

___ 3. Duties of employees working in confined spaces
      ___ *a. Authorized entrant*
           ___ (1) Know the existing and potential confined space hazard(s), the mode(s) of exposure, and the signs, symptoms, and consequences of exposure.
           ___ (2) Use appropriate PPE.
           ___ (3) Maintain communication with attendants to enable them to monitor the entrant's status or alert the entrant to evacuate.

___ (4) Exit from the space in the following situations:

    ___ An authorized person orders the exit.

    ___ The entrant recognizes signs or symptoms of exposure.

    ___ A prohibited condition exists.

    ___ The automatic alarm is activated.

    ___ Alert attendant when

        ___ A prohibited condition exists

        ___ An automatic alarm is activated

___ *b. Attendant*

___ (1) Remain outside permit space during entry operations unless relieved by another attendant.

___ (2) Perform nonentry rescue per procedure, when necessary.

___ (3) Know the existing and potential confined space hazard(s), the mode(s) of exposure, and the signs, symptoms, and consequences of exposure.

___ (4) Maintain communication with and keep an accurate account of those workers entering the permit space.

___ (5) Order evacuation of a permit space when

    ___ An entrant (worker) shows signs of physiological effects of hazard exposure

    ___ An emergency outside the confined space occurs

    ___ The attendant cannot effectively and safely perform the required duties

___ (6) Summon rescue and other services during an emergency.

___ (7) Ensure that unauthorized persons are not allowed in the permit spaces.

___ (8) Ensure that unauthorized persons exit immediately if they have entered the permit space.

___ (9) Inform authorized entrants and entry supervisor of entry by an unauthorized person(s).

___ (10) Perform no duties that could interfere with the above duties.

___ c. *Entry supervisor*

___ (1) Know the existing and potential confined space hazard(s), the mode(s) of exposure, and the signs, symptoms, and consequences of exposure.

___ (2) Verify emergency plans and specified entry conditions such as permits, tests, procedures, and equipment before allowing entry.

___ (3) Terminate entry and cancel permits when entry operations are completed or if a new condition arises.

___ (4) Take appropriate measures to remove unauthorized entrants.

___ (5) Ensure that entry operations remain consistent with the entry permit and that acceptable entry conditions are maintained.

NOTES

F. Fire Protection (Refs. 11, 28, and 29)

___ 1. General

    ___ *a. Terms related to fire protection*

        ___ (1) Fire triangle (See Fig. 4.F.1.)

Figure 4.F.1

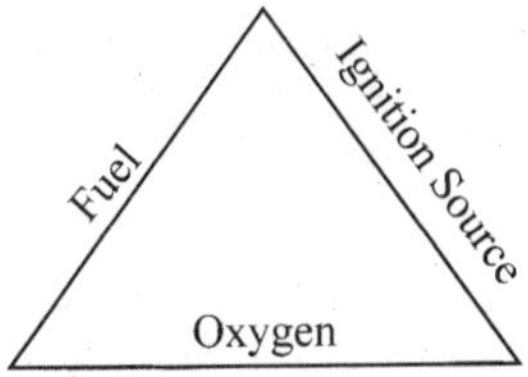

        ___ (2) *Flammable or explosive limits*

        ___ Limits define the concentration range of a combustible material in air at which propagation of flame occurs on contact with an ignition source.

        ___ Limits are expressed in percent vapor or gas in the air by volume.

        ___ Lower flammable limit (LFL) or lower explosive limit (LEL) defines the leanest concentration at which flame propagation will occur.

        ___ Upper flammable limit (UFL) or upper explosive limit (UEL) defines the richest concentration at which flame propagation will occur.

        ___ Examples of limits for selected compounds (See Table 4.F.1.) (Ref. 28).

        ___ (3) *Flash point*

        ___ Defines the lowest temperature at which a liquid gives off enough vapor to form an ignitable mixture with air near the surface of the liquid or within a vessel; and

TABLE 4.F.1   **Flammable Limits for Selected Compounds**

Compound	LFL (% by volume)	UFL (% by volume)
Acetone	2.5	13
Acetylene	2.5	100
1,2-Dichloroethene	9.7	12.8
Gasoline	1.4	7.6
Isobutyl Methyl Ketone	1.4	7.5
Methyl Chloride	10.7	17.4
Propylene	2	11.1
Toluene	1.4	6.7
Xylenes	1	7

_____ The lowest temperature at which flame propagation is capable away from the source of ignition

_____ (4) *Fire point*

    _____ Defines the lowest temperature of a liquid in an open container at which vapors evolve fast enough to support continuous combustion

    _____ Typically a few degrees above the flash point

_____ (5) *Explosion*

    _____ A rapid increase of pressure in a building or container, followed by its sudden release due to rupture.

    _____ The increase in pressure is generally caused by an exothermic chemical reaction or overpressurization of a system.

_____ (5) *Deflagration*

    _____ Propagation of a combustion zone in an unreacted medium at a velocity that is less than the speed of sound.

    _____ Reaction is exothermic.

    _____ Propagation can occur by conduction, convection, and/or radiation.

_____ (6) *Detonation*

    _____ Propagation of a combustion zone in an unreacted medium at a velocity that is greater than the speed of sound.

___ Reaction is exothermic, characterized by the presence of a shock wave.

___ Usually causes an explosion.

___ (7) *BLEVE (boiling liquid expanding vapor explosion)*

___ An explosion caused when fire impinges on the shell of a bulk liquid container, tank, or vessel above the liquid level

___ Causes loss of strength of material

___ Explosive rupture of the structure caused from internal pressure

___ *b.* Classes of fires

___ *Class A*—Fires in ordinary combustible materials such as wood, cloth, paper, rubber, and many plastics

___ *Class B*—Fires in flammable or combustible liquids, flammable gases, greases, oil, paint, and similar materials

___ *Class C*—Fires in live electrical wiring and equipment

___ *Class D*—Fires in certain combustible metals, such as lithium, magnesium, powdered aluminum, potassium, sodium titanium, zinc, and zirconium

___ *Extinguishing agents for Class A, B, C, and D fires* (See Table 4.F.2.)

___ *c.* Sources of ignition

___ (1) Chemical heat energy

___ Heat of combustion—The amount of heat released during a substance's complete oxidation

___ Spontaneous heating—The process whereby a material increases in temperature without drawing heat from its surroundings

TABLE 4.F.2 **Extinguishing Agents for Class A, B, C, and D Fires**

Fire Class	Extinguishing Agents*
A	Water; water solutions that absorb heat and cool, such as foams; certain dry chemicals that coat materials and retard combustion; dry chemicals or halogenated agents that interrupt the combustion chain reaction
B	Dry chemicals that exclude air/oxygen, inhibit the release of vapors, or interrupt the combustion chain reaction; foam agents (excluded if material is water-reactive); carbon dioxide; halogenated agents
C	Nonconductive materials, such as dry chemicals; carbon dioxide; halogenated agents
D	Nonreactive, heat-absorbing materials; combustible metal extinguishing agents such as dry powders, sand, powdered talc, dolomite, zirconium silicate, soda ash, graphite, and sodium chloride

* NOTE: There are direct halon replacement agents and systems that are being phased in for halogentated agents, particularly Halon 1301. These include hydrogen chlorofluorocarbons (HCFCs), hydrogen bromofluorocarbons (HBFCs) and hydrogen fluorocarbons (HFCs).

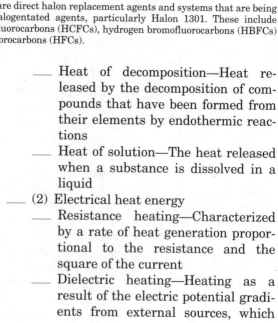

___ Heat of decomposition—Heat released by the decomposition of compounds that have been formed from their elements by endothermic reactions

___ Heat of solution—The heat released when a substance is dissolved in a liquid

___ (2) Electrical heat energy

___ Resistance heating—Characterized by a rate of heat generation proportional to the resistance and the square of the current

___ Dielectric heating—Heating as a result of the electric potential gradients from external sources, which distort the arrangement of the atom; tendency for the electron to move in the direction of the positive potential and for the proton to move in the opposite direction

__ Induction heating—Heating that results from a conductor being subjected to the influence of a fluctuating or alternating magnetic field, or whenever a conductor is in motion across a magnetic field's lines of force

__ Leakage-current heating—Leakage currents (heating) that result because the insulation material is not suited for the service, or the material is too thin

__ Heat from arcing—Heat resulting when an electric circuit carrying current is interrupted; typically severe when motor or other inductive circuits are involved

__ Static electricity heating—Heating resulting from the electrical charge that accumulates on the surfaces of two materials that have been brought together and then separated, with one surface becoming charged positively, the other negatively

__ Heat generated by lightning—Heat generated from the discharge of an electrical charge from a cloud to an opposite charge on the ground

__ (3) Mechanical heat energy

__ Frictional heat—Results from the mechanical energy used in overcoming the resistance to motion when two solids are rubbed together

___ Friction sparks—Sparks that result from the impact of two hard surfaces, at least one of which is usually metal

___ Heat of compression—Heat released when a gas is compressed; also known as the diesel effect

___ (4) Nuclear heat energy

___ Released from the nucleus of an atom when the nucleus is bombarded by energized particles

___ Released in the form of heat, pressure, and nuclear radiation

___ d. Model code organizations and voluntary standards for fire protection

___ National Fire Protection Association, Quincy, Mass.

___ International Conference of Building Officials, Whittier, Calif.

___ Southern Building Code Congress International, Birmingham, Ala.

___ Building Officials and Code Administrators, Country Hills, Ill.

___ American National Standards Institute, New York

___ 2. Hazard ratings

___ a. *NFPA Fire Diamond* (See Fig. 4.F.2.)

___ (1) *Flammability ratings*

___ 4—Extremely flammable

___ 3—Ignites at normal temperatures

___ 2—Ignites when moderately heated

___ 1—Must be preheated to burn

___ 0—Will not burn

___ (2) *Health hazard ratings*

___ 4—Too dangerous to enter

___ 3—Extremely dangerous

___ 2—Hazardous

___ 1—Slightly hazardous

___ 0—Normal material

___ (3) *Reactivity ratings*
    ___ 4—May detonate
    ___ 3—Shock or heat may cause detonation
    ___ 2—Violent chemical change possible
    ___ 1—Unstable if heated
    ___ 0—Normally stable
___ (4) *Special instructions*
    ___ W̶—Avoid use of water
    ___ OX—Oxidizer
    ___ COR—Corrosive
    ___ ACID—Acid
    ___ ALK—Alkali
    ___ Other—Symbols for radioactive materials, biohazard, and other hazards may be used

Figure 4.F.2

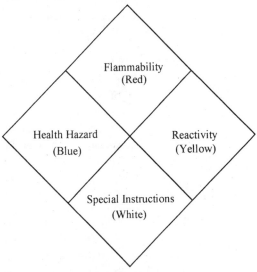

___ (5) *Hazard ratings for selected materials*
(See Table 4.F.3.) (Ref. 28)

TABLE 4.F.3 **Hazard Ratings for Selected Materials**

Hazardous Chemical	Flammability	Health Hazard	Reactivity	Special Instructions
Acetaldehyde	4	2	2	—
Ammonia (gas)	1	2	0	—
Ammonia (liquid)	1	3	0	—
Dibenzene	4	3	3	W̶
Dimethyl Sulfate	1	3	4	—
Dinitrobenzene	1	3	4	—
Hydrogen Cyanide	4	4	2	—
Hydrogen Peroxide	0	2	1	OXY
Phosgene	0	4	0	—
Potassium Perchlorate	0	1	2	OXY
Trichlorosilane	4	3	2	W̶
Trinitrotoluene	4	4	2	—

___ *b.* Uniform fire code classifications of materials according to hazard (Ref. 22)

___ (1) *Liquid and solid oxidizers*

___ Class 4—Can undergo an explosive reaction as a result of contamination or exposure to thermal or physical shock; will enhance the burning rate and may cause spontaneous ignition of combustibles.

___ Class 3—Will cause a severe increase in the burning rate of combustible materials with which it comes in contact or will undergo vigorous self-sustained decomposition resulting from contamination or exposure to heat.

___ Class 2—Will cause a moderate increase in the burning rate or may cause spontaneous ignition of combustible materials with which it comes in contact.

___ Class 1—Will slightly increase the burning rate, but does not cause spontaneous ignition when it comes in contact with combustible materials.

___ (2) *Organic peroxides*

___ Unclassified—Capable of detonation; present an extremely high explosion hazard through rapid explosive decomposition

___ Class I—Capable of deflagration but not detonation; present a high explosion hazard through rapid decomposition

___ Class II—Burn rapidly and present a severe reactivity hazard

___ Class III—Burn rapidly and present a moderate reactivity hazard

___ Class IV—Burn in the same manner as ordinary combustibles and present a minimum reactivity hazard

___ Class V—Do not burn or present a decomposition hazard

___ (3) *Unstable (reactive) materials*

___ Class 4—Readily capable of detonation, or of explosive decomposition or explosive reaction at normal temperatures and pressures due to mechanical or localized thermal shock

___ Class 3—Capable of detonation or of explosive decomposition or explosive reaction, but require a strong initiating source such as thermal or mechanical shock at elevated temperatures and pressures

___ Class 2—Normally unstable and readily undergo rapid release of energy at normal temperatures and pressures; can undergo violent chemical change at elevated temperatures and pressures without detonating

___ Class 1—Normally stable, but can become unstable at elevated temperatures and pressures

___ (4) *Water reactive materials*

___ Class 3—React explosively with water, without heat or confinement

___ Class 2—May form potentially explosive mixtures with water

___ Class 1—May react with water with some release of energy, but not violently

___ 3. Building-planning for fire safety

___ *a.* Means of protecting buildings from fire exposure

___ Automatic sprinkler protection

___ Blank wall of noncombustible materials

___ Barrier wall (self-supporting) between building and exposure

___ Extension of exterior masonry walls to form parapets or wings

___ Automatic outside water curtains for combustible walls

___ Lightning protection systems

___ Emergency and standby power supplies

___ Detection and alarm devices

___ Control of electrostatic ignition sources

___ Good housekeeping practices

___ *b.* Causes of industry and manufacturing fires

___ Processing equipment

___ Spontaneous ignition

___ Static discharges

___ Torches

___ Hand tools

___ Incendiary causes

___ Overcurrent protection devices

___ Heating equipment
___ Smoking materials
___ Cooking equipment
___ Exposure to other hostile fire
___ *c. Life Safety* (See Sec. 4.G.)

NOTES

__ G. Life Safety

__ 1. Building classification of contents (Ref. 31)

    __ *Low-hazard contents*—Those that have such low combustibility that no self-propagating fire can occur

    __ *Ordinary-hazard contents*—Those that are likely to burn with rapidity or to give off a considerable volume of smoke

    __ *High-hazard contents*—Those that are likely to burn with extreme rapidity or explode

__ 2. Building classification of occupancies (Ref. 22)

    __ *a.* Selected examples of general groups

        __ Group A—Assembly

        __ Group B—Business

        __ Group E—Educational

        __ Group F—Factory and industrial

        __ *Group H—Hazardous* (divisions follow)

        __ Group I—Institutional

        __ Group M—Mercantile

        __ Group R—Residential

        __ Group S—Storage

        __ Group U—Utility

    __ *b. Divisions of Group H—hazardous*

        __ Division 1—Occupancies with materials exceeding exempt limits and which present a high explosion hazard

        __ Division 2—Occupancies where combustible dust is manufactured, used, or generated in such a manner that concentrations and conditions create a fire or explosion potential; occupancies with materials exceeding exempt limits, which present a moderate explosion hazard or a hazard from accelerated burning

        __ Division 3—Occupancies where flammable solids, other than combustible dust, are manufactured, used, or generated;

includes materials exceeding exempt limits that present a high physical hazard
___ Division 4—Repair garages not classified as Group S, Division 3 occupancies
___ Division 5—Aircraft repair hangars not classified as Group S, Division 5 occupancies and heliports
___ Division 6—Semiconductor fabrication facilities and comparable research and development areas in which hazardous production materials are used and the aggregate quantity exceeds exempt limits
___ Division 7—Occupancies with materials exceeding exempt limits and which are health hazards

___ 3. Egress in industrial occupancies
___ a. *General purpose occupancy*
___ Contains low- and ordinary-hazard manufacturing operations occurring in any type of building; typically this occupancy has a higher density of employees than other classifications.
___ Travel distance to nearest exit should be 200 ft or less for unsprinklered buildings.
___ If special requirements are met, such as only one story, mandatory automatic sprinkler systems, and smoke and heat venting, the travel distance to nearest exit can be as much as 400 ft.
___ Travel distance to nearest exit should be 250 ft or less for sprinklered buildings.
___ Travel distance to common path of travel should be 50 ft or less for unsprinklered buildings.
___ Travel distance to common path of travel should be 100 ft or less for sprinklered buildings.

___ Dead ends should not be more than 50 ft in length.

___ b. *Special purpose areas*

    ___ Contains low- and ordinary-hazard manufacturing operations; these areas have a low density of employees.

    ___ Travel distance to nearest exit should be 300 ft or less for unsprinklered buildings.

    ___ Travel distance to nearest exit should be 400 ft or less for sprinklered buildings.

    ___ Travel distance to common path of travel should be 50 ft or less for unsprinklered buildings.

    ___ Travel distance to common path of travel should be 100 ft or less for sprinklered buildings.

    ___ Dead ends should not be more than 50 ft in length.

___ c. *High-hazard areas*

    ___ These areas contain processes involving highly combustible, highly flammable, or explosive materials, or materials that are likely to burn with extreme rapidity or to produce poisonous fumes or gases.

    ___ Travel distance to nearest exit should be 75 ft or less for unsprinklered buildings.

    ___ Travel distance to nearest exit should be 75 ft or less for sprinklered buildings.

    ___ Dead ends are not allowed.

    ___ May require addition of exit tunnels, overhead passageways, or horizontal exits.

NOTES

___ H. Construction Safety

___ 1. Subparts of 29 CFR 1926
    ___ *a.* Subpart A—General
        ___ Purpose and scope
        ___ Variances from safety and health standards
        ___ Inspections and right of entry
        ___ Rules of practice for administrative adjudications for enforcement of safety and health standards
    ___ *b.* Subpart B—General interpretations
        ___ Interpretation of statutory terms
        ___ Federal contracts for mixed types of performance
        ___ Relationship to the Service Contract Act and Walsh-Healy Public Act
        ___ Rules of construction
        ___ Other information
    ___ *c.* Subpart C—General safety and health provisions
        ___ Safety training and education
        ___ Recording and reporting of injuries
        ___ First-aid medical attention
        ___ Fire protection and prevention
        ___ Housekeeping
        ___ Illumination
        ___ Sanitation
        ___ PPE
        ___ Acceptable certification
        ___ Shipbuilding and ship repairing
        ___ Incorporation by reference
    ___ *d.* Subpart D—Occupational health and environmental controls
        ___ Medical services and first aid
        ___ Sanitation
        ___ Occupational noise
        ___ Ionizing radiation
        ___ Nonionizing radiation
        ___ Gases, vapors, fumes, dust, and mists

___ Illumination

___ Ventilation

___ Asbestos, tremolite, anthophyllite, and actinolite

___ Hazard communication

___ e. Subpart E—Personal protection and life-saving equipment

___ Head protection

___ Hearing protection

___ Eye and face protection

___ Respiratory protection

___ Safety belts, lifelines, and lanyards

___ Safety nets

___ Working over or near water

___ f. Subpart F—Fire protection and prevention

___ General protection and prevention standards

___ Flammable and combustible liquids

___ Liquefied petroleum gas

___ Temporary heating devices

___ g. Subpart G—Signs, signals, and barricades

___ Accident prevention signs and tags

___ Signaling requirements

___ Barricade requirements

___ h. Subpart H—Materials handling, storage, use, and disposal

___ General requirements for storage

___ Rigging equipment for material handling

___ Disposal of waste materials

___ i. Subpart I—Hand and power tools

___ Requirements for hand tools and power-operated hand tools

___ Abrasive wheels and tools

___ Woodworking tools

___ Lever and ratchet jacks

___ Screw and hydraulic jacks

___ j. Subpart J—Welding and cutting

___ Gas welding and cutting

___ Arc welding
___ Fire prevention
___ Ventilation and protection in welding, cutting

___ *k.* Subpart K—Electrical
___ Installation requirements such as wiring design and protection, wiring methods, components, and equipment for general use
___ Hazardous (classified) locations
___ Safety-related practices such as lockout/tagout
___ Safety-related maintenance of equipment
___ Requirements for special equipment such as battery locations and battery charging

___ *l.* Subpart L—Scaffolding
___ Scaffolding requirements
___ Requirements for guardrails, handrails, and covers

___ *m.* Subpart M—Floor and wall openings
___ Definitions
___ Roof widths

___ *n.* Subpart N—Cranes, derricks, hoists, elevators, and conveyors
___ Equipment requirements for cranes and derricks
___ Helicopters
___ Material hoists, personnel hoists, and elevators
___ Base-mounted drum hoists
___ Overhead hoists
___ Conveyors
___ Aerial lifts

___ *o.* Subpart O—Motor vehicles, mechanized equipment, and marine operations
___ Requirements for motor vehicles
___ Material-handling equipment

    ___ Pile-driving equipment
    ___ Site clearing
    ___ Marine operations and equipment

___ *p.* Subpart P—Excavation
    ___ General requirements
    ___ Requirements for protective systems
    ___ Soil classification
    ___ Sloping and benching
    ___ Shoring for trenches

___ *q.* Subpart Q—Concrete and masonry construction
    ___ Requirements for equipment and tools
    ___ Requirements for cast-in-place and precast concrete
    ___ Requirements for lift-slab construction operations
    ___ Requirements for masonry construction

___ *r.* Subpart R—Steel erection
    ___ Flooring requirements
    ___ Bolting, riveting, fitting-up, and plumbing-up

___ *s.* Subpart S—Underground construction, caissons, cofferdams
    ___ Underground construction and caissons
    ___ Cofferdams
    ___ Compressed air

___ *t.* Subpart T—Demolition
    ___ Preparatory operations
    ___ Stairs, passageways, and ladders
    ___ Chutes
    ___ Removal of materials through floor openings
    ___ Removal of walls, masonry, sections, and chimneys
    ___ Removal of floors and walls
    ___ Removal of steel construction
    ___ Mechanical demolition
    ___ Selective demolition by explosives

___ *u.* Subpart U—Blasting and use of explosives
    ___ Blaster qualifications
    ___ Surface and underground transportation of explosives
    ___ Storage of explosives and blasting agents
    ___ Loading of explosives or blasting agents
    ___ Initiation of explosive charges—electric blasting

___ *v.* Subpart V—Power transmission and distribution
    ___ Tools and protective equipment
    ___ Mechanical equipment
    ___ Material handling
    ___ Grounding for protection of employees
    ___ Overhead and underground lines
    ___ Construction in energized substations
    ___ External load helicopters
    ___ Linesman's body belts, safety straps, and lanyards

___ *w.* Subpart W—Roll-over protective structures; overhead protection
    ___ Roll-over protective structures for material-handling equipment
    ___ Minimum performance criteria for roll-over protective structures for designated scrapers, loaders, dozers, graders, and crawler tractors
    ___ Protective frame test procedures and performance requirements for wheel-type agricultural and industrial tractors used in construction

___ *x.* Subpart X—Stairways and ladders
    ___ Requirements for stairways and ladders
    ___ Training requirements

___ 2. *Construction safety training checklist* (See Table 4.H.1.)

___ 3. *Material storage requirements checklist* (See Table 4.H.2.)

___ 4. *Scaffolding requirements checklist* (See Table 4.H.3.)
___ 5. *Sling inspection checklist* (See Table 4.H.4.)

TABLE 4.H.1   Construction Safety Training Checklist

Supervisor Training Topics

- Safety plan
- Emergency plan
- Selection and use of PPE for project tasks
- Safe work practices
- Schedule for testing safety equipment
- Accident investigation techniques
- Qualifications/training of employees
- Recordkeeping requirements
- Rights of employees under OSHA

Employee Training Topics

- Safety plan
- Emergency plan
- Hazard communication
- Safe use of PPE
- Confined space entry
- Scaffolding
- Perimeter guarding
- Housekeeping
- Material handling
- Rigging
- Crane safety
- Electrical safety
- Trenching and excavation

Weekly Safety Meeting Topics

- Safe working conditions
- Safe use of tools
- Safe lifting techniques
- Incidents and near misses
- Specific information about job tasks

TABLE 4.H.2   **Material Storage Requirements Checklist**
(Ref. 2)

General Requirements

- Are all materials stored in tiers stacked, racked, blocked, inter-locked, or otherwise secured to prevent sliding, falling, or collapse?
- Are maximum safe-load limits of floors within buildings and structures, in pounds per square foot, conspicuously posted in all storage areas, except for floor or slab on grade?
- Are maximum safe loads observed at all times?
- Are aisles and passageways kept clear to provide for the free and safe movement of material-handling equipment or employees?
- Are aisles and passageways in good repair?
- When a difference in road or working levels exists, are there ramps, blocking, grading, or other means used to ensure the safe movement of vehicles between the two levels?

Material Storage Requirements

- Is material stored inside buildings under construction placed more than 6 feet from any hoistway or inside floor openings, and is it more than 10 feet from an exterior wall that does not extend above the top of the material stored?
- Are employees required to be equipped with lifelines and safety belts while working on stored material in silos, hoppers, tanks, and similar storage areas?
- Are noncombustible materials segregated in storage?
- Are bagged materials stacked by stepping back the layers and cross-keying the bags at least at the level of every tenth bag?
- Are materials removed from scaffolds or runways except for supplies needed for immediate operations?
- Are brick stacks 7 feet or less in height?
- When a loose brick stack reaches a height of 4 feet, is it tapered back 2 inches for every foot above the 4-foot level?
- When masonry blocks are stacked higher than 6 feet, is the stack tapered back one-half block per tier above the 6-foot level?
- Does used lumber have all nails withdrawn before stacking?
- Is lumber stacked on level and solidly supported sills, and is lumber stacked so as to be stable and self-supporting?
- Are lumber piles 20 feet or less in height?
- Is lumber to be handled manually stacked 16 feet high or less?
- Are cylindrical materials, such as structural steel, poles, pipe, and bar stock, stacked and blocked so they do not spread or tilt?

Housekeeping

- Are storage areas kept free from accumulation of materials that constitute hazards from tripping, fire, explosion, or pest infestation?
- Is vegetation control exercised when necessary?

TABLE 4.H.2   **Material Storage Requirements Checklist**
(Ref. 2) (*Continued*)

Dockboards (bridge plates)

- Are portable and powered dockboards strong enough to carry the load imposed on them?
- Are portable dockboards secured in position, by being either anchored or equipped with devices which will prevent their slipping?
- Are handholds or other effective means provided on portable dockboards to permit safe handling?
- Is positive protection provided to prevent railroad cars from being moved while dockboards or bridge plates are in position?

TABLE 4.H.3   **Scaffolding Requirements Checklist** (Ref. 3)

- Does scaffolding have sound footing/anchorage capable of carrying maximum load capabilities?
- Are erection, movement, alteration, or dismantlement of scaffolding performed under the supervision of competent persons?
- Are guardrails and toeboards on open ends and sides of scaffolds more than 10 feet above the ground?
- Do scaffolds that are 4 to 10 feet in height and have a minimum horizontal dimension in either direction of less than 45 inches have standard guardrails on all open sides and end of the platform?
- Are guardrails 2 × 4 inches or equivalent, approximately 42 inches high, with a midrail, when required?
- Are supports located at intervals 8 feet or less in height?
- Are toeboards a minimum of 4 inches in height?
- If persons can pass or work under the scaffold, are there screened guardrails between the midrail and the toeboard?
- Are damaged parts of the scaffold repaired or replaced immediately?
- Are load carrying timber members of the scaffold framing a minimum of 1500 fiber (stress grade) construction grade lumber?
- Is planking scaffold grade or equivalent?
- Does the scaffold design follow the permissible spans for 2- by 10-inch or wider planks?
- Is the maximum permissible span for 1¼ × 9 inches or wider of full thickness 4 feet with medium duty loading of 50 pounds per square foot?
- Do scaffold planks extend over their end supports not less than 6 inches or more than 12 inches?
- Is an access ladder or equivalent safe access provided?
- Are scaffold poles, legs, or uprights plumb, secured, and rigid?
- When there are overhead hazards, do workers wear hard hats and is there overhead protection?

TABLE 4.H.3 **Scaffolding Requirements Checklist** (Ref. 3) (*Continued*)

- Are slippery hazards eliminated immediately?
- Is welding, burning, riveting, or open flame work prohibited on any staging suspended by fiber or synthetic rope?
- Are wire, synthetic, or fiber rope used for scaffold support capable of supporting at least 6 times the rated load?
- Is use of shore or lean-to scaffolds prohibited?
- Do materials being hoisted onto a scaffold have a tag line?
- Are employees prohibited from work on scaffold during storms or high winds?
- Are tools, materials, and debris put away promptly after use to avoid becoming a hazard?

TABLE 4.H.4. **Sling Inspection Checklist**

Wire Rope Inspections—Look for

- Randomly distributed broken wires in one rope lay
- Wear or scraping of outside individual wires
- Kinking of the rope
- Crushing of the rope
- *Bird caging,* or the twisting or distending of a strand or strands in a multistrand rope, due to torsion
- *High stranding,* or the raising of one strand of the rope
- Bulges in rope
- Gaps or excessive clearance between strands
- Core protrusion
- Heat damage, torch burns, or electric arc strikes
- Unbalanced severely worn area
- Cracks, deformities, wearing, or corrosion of the end attachments
- Stretched or twisted hook-throat opening

Chain Sling Inspections—Look for

- Localized stretch or wear
- Stretch throughout the entire length of chain
- Grooving
- Twisted or bent links
- Cracks
- Gouges or nicks
- Corrosion pits
- Burns
- Integrity of master links and hooks
- Hook-throat opening is not stretched or twisted

NOTES

References

NOTES

1. Code of Federal Regulations, 29 CFR 1910.95(a), Table G-16.
2. Code of Federal Regulations, 29 CFR 1926.250.
3. Code of Federal Regulations, 29 CFR 1926.451.
4. Code of Federal Regulations, 40 CFR 131.36(b)(1).
5. Code of Federal Regulations, 40 CFR Part 403.
6. Code of Federal Regulations, 40 CFR Part 60, Appendix A.
7. American Conference of Governmental Industrial Hygienists (ACGIH), *Guide to Occupational Explosive Values–1996,* Cincinnati, Ohio, 1996.
8. American Conference of Governmental Industrial Hygienists (ACGIH), *Industrial Ventilation, A Manual of Recommended Practice,* Cincinnati, Ohio, 1992.
9. American Conference of Governmental Industrial Hygienists (ACGIH), *TLVs and BEIs and Guide to Occupational Exposure—1997,* Cincinnati, Ohio, 1997.
10. American Industrial Hygiene Association, *Odor Thresholds for Chemicals with Established Occupational Health Standards,* Fairfax, Va., 1989.
11. American Society of Safety Engineers (ASSE), *Dictionary of Terms Used in the Safety Profession,* Des Plaines, Ill., 1988.
12. American Society for Testing and Materials (ASTM), *Environmental Site Assessments—Phase I: Assessment, Process, Practice,* E 1527, Philadelphia, Pa., 1994.
13. Centers for Disease Control and Prevention, *Fatal Injuries to Workers in the United States, 1980–1989: A Decade of Surveillance,* DHHS (NIOSH) Number 93-108, U.S. Department of Health and Human Services, Public Health Service, Cincinnati, Ohio, 1993.
14. Drucker, Peter F., *Toward the Next Economics and Other Essays,* Harper & Row, New York, 1981.
15. Fiksel, Joseph, *Design for Environment,* McGraw-Hill, New York, 1996.
16. Forsberg, Krister, and S. Z. Mansdorf, *Quick Selection Guide to Chemical Protective Clothing,* Van Nostrand Reinhold, New York, 1993.

17. Freeman, Harry M., "Developing and Maintaining a Pollution Prevention Program," in *Industrial Pollution Prevention Handbook,* McGraw-Hill, New York, 1995.
18. Freeman, Harry M., "Overview of Waste Reduction Techniques Leading to Pollution Prevention," in *Industrial Pollution Prevention Handbook,* McGraw-Hill, New York, 1995.
19. Grainger Industrial and Commercial Equipment and Supplies, 1997.
20. Hance, B. J., Caron Chess, and Peter M. Sandman, *Industry Risk Communication Manual: Improving Dialogue with Communities,* Lewis Publishers, Boca Raton, Fla., 1990.
21. Herzberg, Frederick, *Work and the Nature of Man,* World Publishing Company, Cleveland, Ohio, 1966.
22. International Conference of Building Officials, *Uniform Building Code,* Whittier, Calif., 1994.
23. Karr, Al, "Top 10 OSHA Violations," *Safety & Health,* December 1997, pp. 46–50.
24. Lewis, Richard, J., *Hawley's Condensed Chemical Dictionary,* 12th ed., Van Nostrand Reinhold, N.Y., 1993.
25. Maslow, A. H., *Motivation and Personality,* Harper & Brothers, New York, 1954.
26. McGregor, Douglas, *The Human Side of Enterprise,* McGraw-Hill, New York, 1960.
27. Means, R. S., *Construction Cost Estimating,* 1997.
28. National Fire Protection Association, *Fire Protection Guide on Hazardous Materials,* Quincy, Mass., 1991.
29. National Fire Protection Association, *Fire Protection Handbook,* Quincy, Mass., 1997.
30. National Fire Protection Association, *Hazardous Materials Response Handbook,* Quincy, Mass., 1989.
31. National Fire Protection Association, Life Safety Code®, NFPA 101, Quincy, Mass., 1994.
32. National Safety Council (NSC), *Accident Prevention Manual for Business and Industry,* 10th ed., Chicago, Ill., 1992.

33. National Safety Council (NSC), *Fundamentals for Industrial Hygiene,* 3d. ed., Chicago, Ill., 1988.

34. Occupational Safety and Health Association (OSHA), *Concepts and Techniques of Machine Guarding,* PB93-129930, Washington, D.C., 1992.

35. Schweitzer, Philip A., *Corrosion and Corrosion Protection Handbook,* 2d ed., Marcel Dekker, New York, 1989.

36. Talty, John T., *Industrial Hygiene Engineering,* Noyes Data Corporation, Park Ridge, N.J., 1988.

37. U.S. Environmental Protection Agency, *Compendium of Methods for the Determination of Toxic Organic Compounds in Ambient Air,* Research Triangle Park, N.C., 1988.

38. U.S. Environmental Protection Agency, *Compendium of Technologies Used in the Treatment of Hazardous Waste,* Cincinnati, Ohio, 1987.

39. U.S. Environmental Protection Agency, *Handbook of Suggested Practices for the Design and Installation of Ground-Water Monitoring Wells,* Environmental Monitoring Systems Laboratory, Las Vegas, Nev., 1990.

40. U.S. Environmental Protection Agency, *RCRA Ground-Water Monitoring: Draft Technical Guidance,* Washington, D.C., 1992.

41. U.S. Environmental Protection Agency, *Storm Water Management and Technology,* Noyes Data Corporation, Park Ridge, N.J., 1993.

42. Woodside, Gayle, *Hazardous Materials and Hazardous Waste Management,* John Wiley & Sons, New York, 1993.

43. Woodside and Kocurek, *Environmental, Safety, and Health Engineering,* Wiley & Sons, New York, 1997.

44. Woodside, Gayle, Patrick Aurrichio, and Jeanne Yturri, *ISO 14001 Implementation Manual,* McGraw-Hill, New York, 1998.

NOTES

Index

NOTES

A

B

C

NOTES

ABOUT THE AUTHOR

Gayle Woodside is a program manager on IBM's Corporate Environmental Affairs staff. She has over 15 years of experience in chemical management, environmental treatment systems design and operation, process safety, environmental permitting, and ISO 14000. A professional engineer and a Certified Safety Professional, she is co-author of McGraw-Hill's *ISO 14000 Guide,* and *ISO 14001 Implementation Manual.*

NOTES

NOTES

NOTES

NOTES

NOTES

NOTES

NOTES